SpringerBriefs in Applied Sciences and Technology

For further volumes:
http://www.springer.com/series/8884

Nagamitsu Yoshimura

Historical Evolution Toward Achieving Ultrahigh Vacuum in JEOL Electron Microscopes

 Springer

Dr. Nagamitsu Yoshimura
3-22-75 Fujimoto, Kokubunji
Tokyo 185-0031, Japan

ISSN 2191-530X   ISSN 2191-5318 (electronic)
ISBN 978-4-431-54447-0   ISBN 978-4-431-54448-7 (eBook)
DOI 10.1007/978-4-431-54448-7
Springer Tokyo Heidelberg New York Dordrecht London

Library of Congress Control Number: 2013948408

Printed on acid-free paper

Springer is part of Springer Science+Business Media (www.springer.com)

Preface

This book describes the history of how ultrahigh-vacuum (UHV) JEOL electron microscopes (JEM series) were brought into existence, from their conception to the birth of the electron microscope.

My co-workers and I engaged in developing vacuum technology for electron microscopes at JEOL (Japan Electron Optics Laboratory Co. Ltd.) for many years. This book now presents the UHV technology of JEMs.

The column of the electron microscope (EM), including the camera chamber, is very complex in construction and is composed of many units and parts.

A high-tension (HT) electron gun is installed on the top of the column for illuminating the specimen with a fine electron probe. A very clean vacuum is required for preventing contamination build-up on the specimen surface. When a microdischarge occurs on the HT electron gun, noiseless EM-image observation becomes impossible.

A very clean UHV around the specimen is essential. However, some types of UHV pumps, such as the turbo-molecular pump (TMP) and the cryopump (CP), cause vibration, making noiseless EM-image observation impossible.

We had to develop two types of sputter ion pumps (SIPs) for creating UHV-EMs, one having high pumping speeds in the UHV region, and the other having stable pumping performance for inert gases such as argon for thinning specimens, in addition to their demonstrating high performance in the UHV region. JEOL SIPs for our electron microscopes are described in detail in this book.

The users of EMs are high-level researchers, working at the frontiers of new materials or new biological specimens. They often use the EM under extreme conditions, with problems sometimes occurring in the vacuum system of the users' EMs. We must resolve such problems as quickly as possible and improve the vacuum system in order to prevent the recurrence of such problems. Typical examples of users' claims are presented in this book, such as microdischarge occurring on an electron gun and backstreaming of diffusion pump (DP) oil vapor. Accidents occurring in users' EMs showed us how to improve the vacuum systems of JEMs.

We sincerely hope that this short book will be read by many engineers and researchers using analytical instruments that employ a fine electron probe, such as the EM, the X-ray microanalyzer (XMA), and the Auger electron spectrometer (AES).

Tokyo, Japan   Nagamitsu Yoshimura

Acknowledgments

I would like to express my deep appreciation to Dr. Y. Harada of JEOL Ltd. for his cooperation and suggestions when I was writing this book. I greatly appreciate JEOL's permission to present some parts of the descriptions from the History of JEOL(1): 35 years from its birth (March 1986) and the History of JEOL(2): 60 years from its birth, both of which were romantic books edited by JEOL Co. Ltd. written in Japanese. I also thank JEOL for permitting the sharp photographs of their first microscope, the DA-1, be presented in Fig. 2.2 of this book.

In addition, I am grateful to JEOL for permission to present its expert technologies based on our experiments.

Due thanks are given to colleagues at JEOL Ltd., especially to Mr. H. Hirano. Knowing that we all did our best in working on the vacuum technology for electron microscopes is indeed a nice memory.

Contents

Chapter 1
Introduction of the Electron Microscope

Abstract This short book mainly describes the ultrahigh-vacuum (UHV) technology in JEOL electron microscopes.

There are two types of electron microscopes, one the transmission electron microscope (TEM) and the other scanning electron microscope, accurately speaking, scanning secondary-electron microscope (SEM). From the view point of vacuum technology, it is much difficult to evacuate the complex column of TEM to UHV, compared in the case of SEM. Therefore, we, the engineers engaging in vacuum technology, mainly made efforts in improving the vacuum in TEMs. At the same time, we naturally engaged in improving the vacuum systems of SEM and other scientific instruments such as X-ray microanalyzer (XMA) and Auger-electron spectrometer (AES). Additionally speaking, SEM equipped with the field emission (FE) electron source earlier than TEM did, and AES was a UHV instrument from its birth. The large sputter ion pump (SIP) for the AES chamber must pump Argon gas stably for depth-profile analysis.

The TEM has been developing to the analytical EM (AEM) by having an X-ray analysis function, and to the scanning EM (STEM) by having a scanning function.

It would be better to introduce the principle of EM and the construction of the microscope column briefly. The principles SEM, XMA, and AES are also presented briefly.

1.1 Principle

1.1.1 Signals Coming from the Specimen Illuminated with a Fine Electron Beam

When a narrow electron beam illuminates the specimen, many signals with information are obtained, as shown in Fig. 1.1.

N. Yoshimura, *Historical Evolution Toward Achieving Ultrahigh Vacuum in JEOL Electron Microscopes*, SpringerBriefs in Applied Sciences and Technology, DOI 10.1007/978-4-431-54448-7_1, © The Author(s) 2014

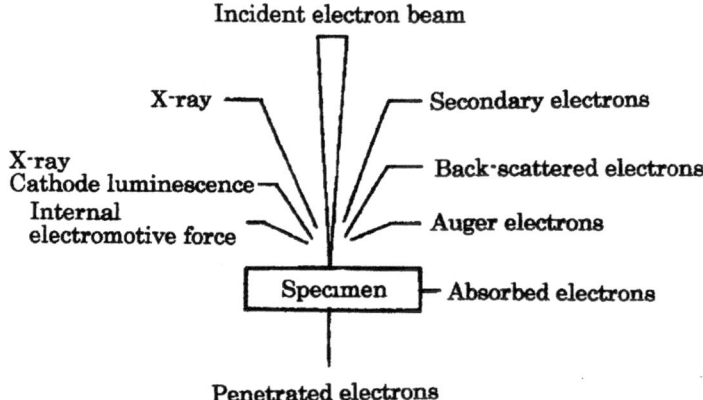

Fig. 1.1 Many signals with information from the specimen when illuminated with a narrow electron beam [1]

An electron beam emitted from the hot filament of the electron gun is accelerated by the voltage between cathode and anode, and then focused by the condenser lens, thus irradiating the specimen. The transmitted electron beam, which has the information on the specimen construction, enlarged by the image-formation lens system. Besides, secondary electrons, backscattered electrons, and characteristic X-rays, which have respective information, are generated. The transmitted electrons receive an energy loss depending on the excitation of the compositing elements. Detecting the amount of such energy loss makes analyzing combination state of the constituting elements possible.

The energy dispersive spectrometer (EDS), attached to the electron microscope, gives the microscope the function of electron probe micro-analyzer (AEM). The scanning transmission electron microscope (STEM) can give the STEM images.

1.1.2   Transmission Electron Microscope

Transmission electron microscopy (TEM) is a microscopy technique whereby a beam of electrons is transmitted through an ultra-thin specimen, interacting with the specimen as it passes through. An image is formed from the interaction of the electrons transmitted through the specimen; the image is magnified and focused onto an imaging device, such as a fluorescent screen, on a layer of photographic film, or to be detected by a sensor such as a CCD camera.

The first TEM was built by Max Knoll and Ernst Ruska in 1931, with this group developing the first TEM with resolving power greater than that of light in 1933 and the first commercial TEM in 1939.

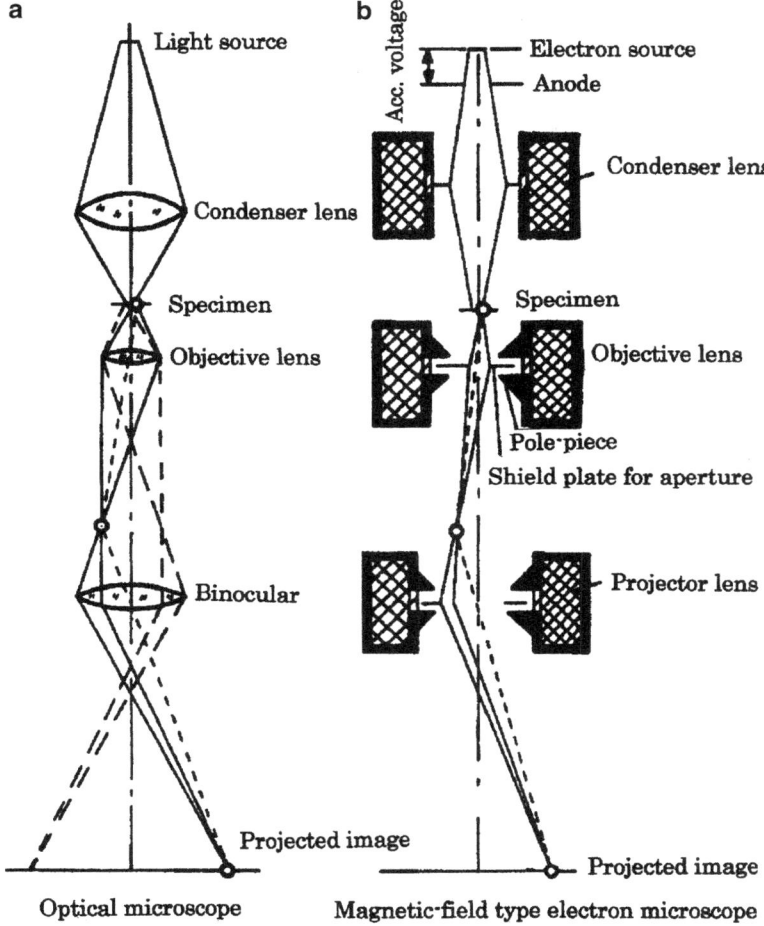

Fig. 1.2 Ray paths for optical microscope (**a**) and electron beam paths for transmission electron microscope (TEM) (**b**) [1]

The principle of image formation by electron lenses is analogous to that by optical lenses. Electron lenses have five Seidel aberrations (spherical aberration, coma, astigmatism, curvature of field, and distortion), which are almost same as optical lenses.

Electron lenses can give the convex lens only, whose focal length varies widely depending on the coil current (Fig. 1.2).

Figure 1.3 shows electron beam paths for diffraction image formation, which is compared with the electron beam paths for transmission electron image formation. As seen in Fig. 1.3, the diffraction image is formed on the back focal plane of the objective lens. One can see the diffraction image on the screen when making the back focal plane of the objective lens be moved to the screen plane by adjusting the focal length of the intermediate lens.

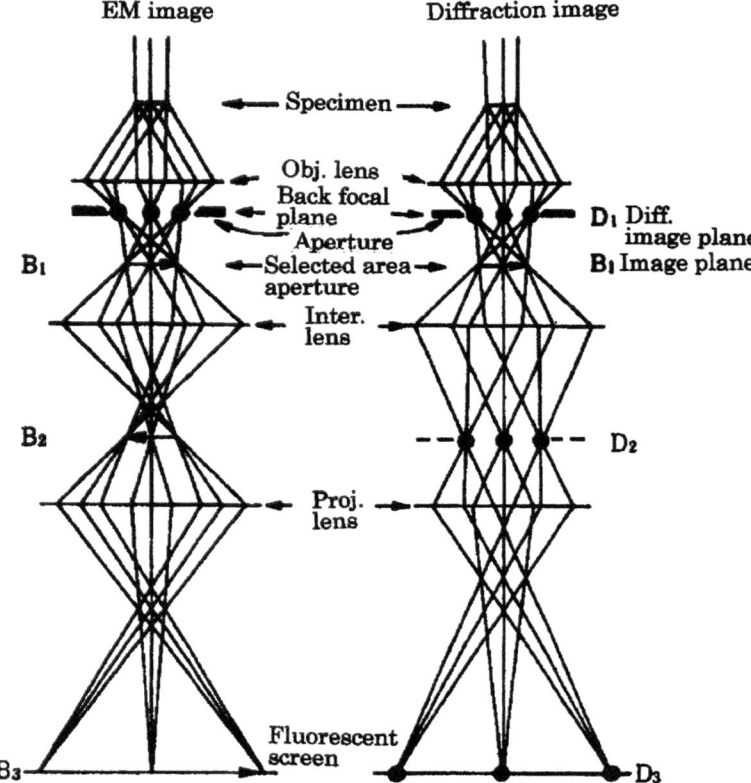

Fig. 1.3 Formation of the diffraction image, compared with that of transmission electron image [1]

1.1.3   Scanning Electron Microscope

A scanning electron microscope (SEM) is a type of electron microscope that produces images of a sample by scanning it with a focused beam of electrons. The electrons interact with electrons in the sample, producing various signals that can be detected and that contain information about the sample's surface topography and composition. The electron beam is generally scanned in a raster scan pattern, and the beam's position is combined with the detected signal to produce an image. SEM can achieve resolution better than 1 nm. Specimens can be observed in high vacuum, low vacuum, and in environmental SEM specimens can be observed in wet condition.

The principle of formation of scanning secondary electron image is similar to that of TV-image formation, which is shown in Fig. 1.4.

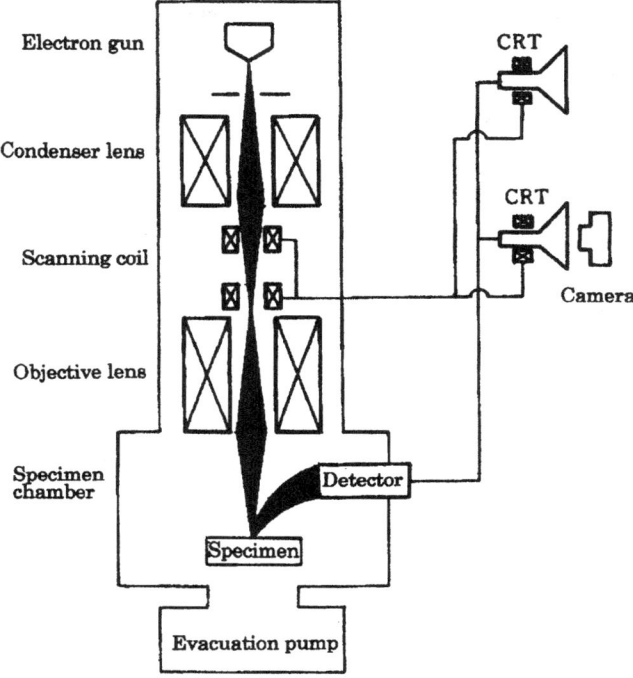

Fig. 1.4 Principle of formation of scanning secondary electron image of SEM [1]

The image contrast is obtained by tilting angle effect and edge effect, as shown in Figs. 1.5 and 1.6. Due to these effects one can observe the stereograph image using the SEM.

1.1.4   X-Ray Microanalyzer

An X-ray microanalyzer (XMA), also known as an electron probe microanalyzer (EPMA), is an analytical tool used to non-destructively determine the chemical composition of small volumes of solid materials. It works similarly to a SEM: the sample is bombarded with an electron beam, emitting X-rays at wavelengths characteristic to the elements being analyzed. This enables the abundances of elements present within small sample volumes (typically 10–30 cubic micrometers or less) to be determined. The concentrations of elements from boron to plutonium can be measured at levels as low as 100 parts per million (ppm).

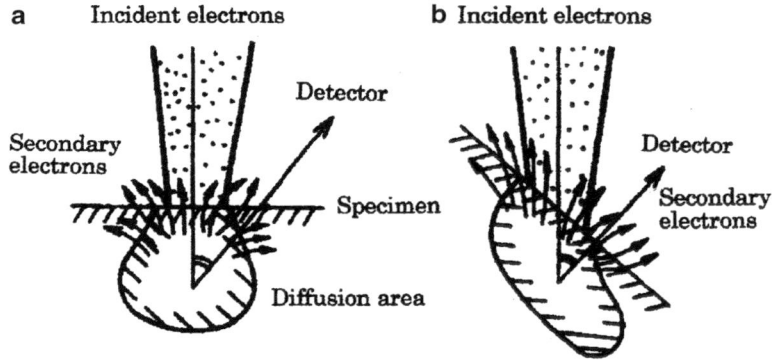

Fig. 1.5 Tilting angle effect in SEM [1]

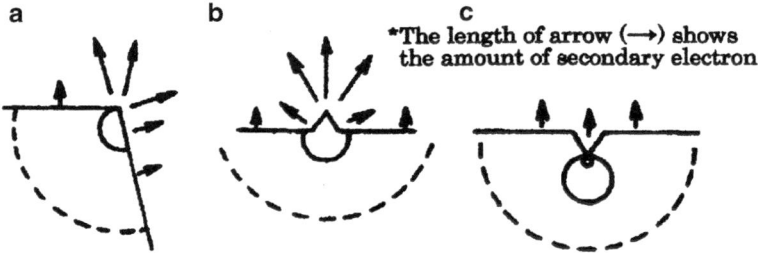

Fig. 1.6 Edge effect in SEM [1]

Two types of detection methods for characteristic X-ray are there, WDS (wavelength-dispersive X-ray spectroscopy) method and EDS (energy-dispersive X-ray spectroscopy) method. Illustration of WDS and EDS systems are presented in Fig. 1.7.

Analytical electron microscopes (AEMs) generally provided with EDS (energy-dispersive X-ray spectroscope).

1.1.5 Auger Electron Spectrometer

Auger electron spectroscopy (AES) is a common analytical technique used specifically in the study of surfaces and, more generally, in the area of materials science. Underlying the spectroscopic technique is the Auger effect, as it has come to be called, which is based on the analysis of energetic electrons emitted from an excited atom after a series of internal relaxation events.

When the sample surface is illuminated Auger electrons are emitted with Auger energy. Auger energy has the characteristic energy of the elements composing the surface, making identification of the surface elements from its energy spectrum. Illustration of AES system is presented in Fig. 1.8.

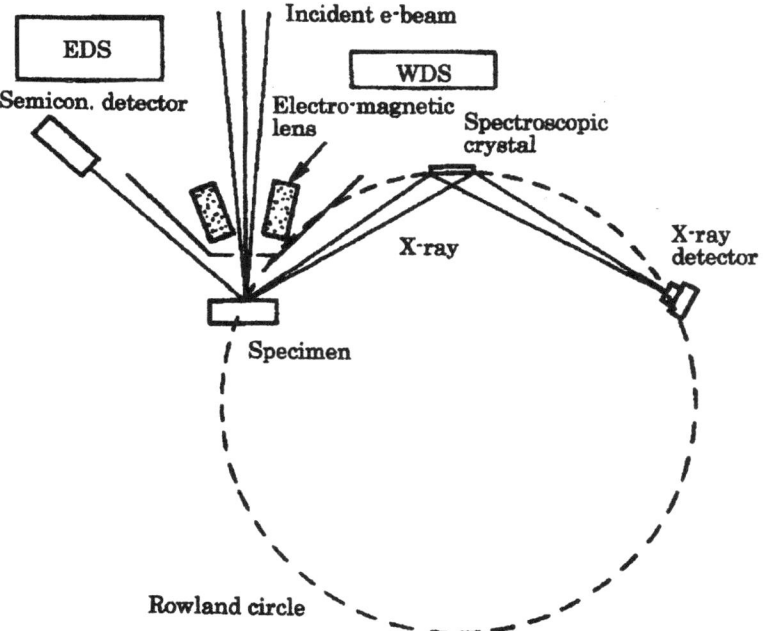

Fig. 1.7 Sketch of WDS and EDS [1]

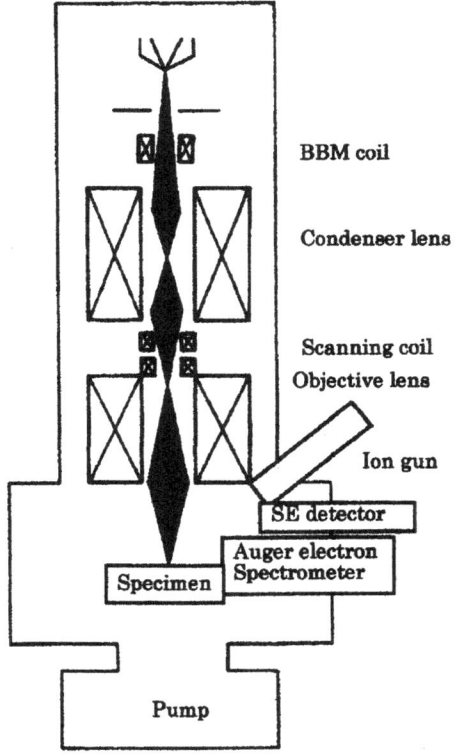

Fig. 1.8 Illustration of AES system [1]

1.2 Construction of EM Column

Clean vacuum in the column was essential in order to minimize specimen contamination build-up.

Figure 1.9 shows the crosssection of the typical electron microscope, JEM-1200EX (120 kV). A specimen is inserted between the objective lens pole-pieces, located just below the specimen chamber. The electron emitter of W-filament is located inside the Wehnelt electrode of the electron gun. Many sheets of electron sensitive film are inside the dispensing magazine and the receiving magazine located in the camera chamber. The viewing chamber has a large volume of nearly 40 L.

The whole column is composed of the gun chamber, the condenser-lens system, the specimen chamber (or called as the mini-lab. chamber), the goniometer stage, the objective lens/pole-pieces system composed of objective lens, the intermediate

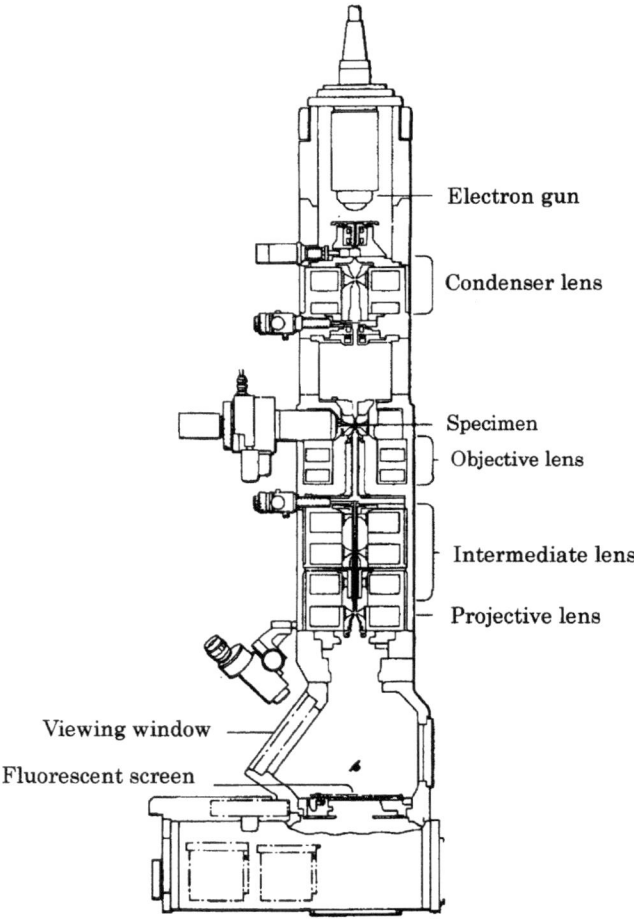

Fig. 1.9 Crosssection of JEM-1200EX (120 kV-EM) [1]

lens and the projector lens, the viewing chamber and the camera chamber. The viewing chamber including the camera chamber is often called as the viewing chamber or camera chamber.

The electron beam from the W-filament of the electron gun is accelerated by the anode voltage, and focused by the condenser-lens system to illuminate the specimen. The transmitted electron beam, having the information of specimen construction, is spread by the image-formation lens system (objective lens/intermediate lens/projector lens systems), displaying an EM-image on the fluorescent screen inside the viewing chamber. When the specimen is illuminated with an electron beam, secondary electrons, backscattered electrons and X-rays are generated with different information respectively, besides transmitted electrons. The transmitted electrons also have received the energy loss through compositional elements of the specimen. By detecting such signals the elemental composition and the coupling state of elements can be analyzed.

A W-filament is positioned inside the Wehnelt electrode of the electron gun. The hole of anode electrode is facing to the hole of Wehnelt electrode. The filament and the Wehnelt electrode is kept at negative high-voltage (for instance, -100 kV), being insulated from the earth potential. There is an air-lock valve beneath the gun chamber, making the column being evacuated during the period for exchanging the filament in the vented gun chamber.

The beam path of the condenser-lens system is so narrow, showing high impedance for gas flow. The condenser-lens aperture assembly can be removed and inserted manually. There is the specimen chamber (often called as the mini-lab. chamber) under the condenser-lens system, to which the pumping pipe with a relatively large diameter is attached. The goniometer stage having Z-control function and the objective-lens pole-pieces are both positioned just below the specimen chamber. After the specimen holder is inserted into the goniometer stage, the specimen space is roughly evacuated, then the airlock valve inside the goniometer is opened and the specimen at the end of the specimen holder is inserted between the objective-lens pole pieces. A wide-bore evacuation pipe is attached for the pole pieces in order to minimize contamination build-up.

The beam path of the image-formation lens system (objective/intermediate/projector lens systems) is also so narrow showing high impedance for gas flow. Notice that a small aperture of about 0.6 mm diameter is positioned at the under bore of the projector-lens pole pieces.

The dispensing magazine and the receiving magazine inside the camera chamber contain many sheets of film. The air-lock valve for the camera chamber is located between the projector-lens system and the viewing chamber, making the dispensing magazine and the receiving magazine be exchanged while the column being evacuated in high vacuum.

Next, let us see the composition material of the column. The walls of the gun chamber and the microscope column are made of steel in order to shield the external electro-magnetic wave from the electron beam and to shield the X-ray generated when the metal surface is irradiated with high-speed electrons. The viewing chamber and the camera chamber are made of brass or steel casting by the reasons that they are complex in shape and their walls must shield the generating X-ray. The pole pieces of every magnetic lens are made of pure steel. On the other hand, the specimen stage and the specimen holders are made of non-magnetic materials like aluminum, titanium or copper. The aperture plates are made of Ta because their temperatures become very high when irradiated with high-speed electrons. Parts of porcelain or ceramics are used for the electron gun and the electron detectors. The O-ring seals are generally Viton-A (fluorine elastic material). For Viton-A O-rings at the movable parts a very small amount of lubricant grease is generally applied. An adhesive agent of organic material is also applied for some composition parts. Therefore, the microscope column cannot be baked at a high temperature of about 100 °C.

Large-size magnetic lenses are located out of vacuum. Recently, small-size deflection coils are located out of vacuum for minimizing the contamination build-up.

As seen above, it can be said that the whole column of EM is complex in construction, composed of a large number of mechanical parts. It is indeed very difficult to achieve a clean, ultrahigh vacuum (UHV) in the specimen chamber under such conditions.

1.3 Vacuum Quality and Pressure Required for EM

Vacuum pressures required are generally as follows.

1. In order to prevent scattering of the electron beam due to colliding with residual gas molecules, a high vacuum of about 10^{-3} Pa is required.
2. In order to prevent burning out of the hot W-filament or the LaB_6 emitter, a high vacuum is required. Partial pressures of O_2 and H_2O are desired to be sufficiently low. Generally speaking, ~10^{-4} Pa for a W-filament, ~10^{-5} Pa for a LaB_6 emitter, and ~10^{-8} Pa for an FE emitter, are required.
3. In order to prevent the microdischarge on the electron gun or the accelerating tube, a high vacuum of 10^{-4}–10^{-5} Pa is required.
4. EM hates any vibration of the column or key units like the specimen stage or the electron gun for observing high-resolution EM images. In order to prevent vibration, any vacuum pump is required not to cause such vibration. Sound emitted from vacuum pumps may cause vibration, so a silent pump is desired.
5. Very clean high vacuum is required in the vicinity of the specimen. This requirement is achieved by pumping the gas molecules around the specimen effectively by using the effective anti-contamination device (ACD) cooled with liq. nitrogen (N_2).

Among these requirements the last one accompanies practical difficulty to achieve.

Reference

1. JEOL (1986) History of JEOL(1); 35 years from its birth (in Japanese), JEOL Co. Ltd. March 1986

Chapter 2
History of JEOL Electron Microscopes

Abstract It could be said that the electron microscope was first invented by Knoll and Ruska (in Germany). Borries and Ruska (Siemens-Halske) developed the 100 kV EM 1932, whose resolution was 10 nm, which would be the first birth of the electron microscope in the world. About 10 years later Hitachi and Toshiba in Japan developed the electron microscope, respectively 1941.

JEOL challenged developing EM, a little bit later. The birth of AD-1 EM, assembled on a wood table, was the debut of our company, JEOL. It was September 1947, a few years after the end of the Pacific War.

2.1 First 20 Years from the Invention of Electron Microscope in the World

The supplement of the journal, *Microscopy* (written in Japanese), was issued titled "Development of Technology of Electron Microscope in Japan" 2011 [1], where the course of history of the development of the EM technology in Japan is described in details.

Figure 2.1 shows the history around EM in the world 1932–1952 [1]. Knoll and Ruska (in Germany) first invented the EM 1932 and submitted the monograph titled "Das Electronenmikroskop" [2]. And, Borries and Ruska (Siemens-Halske) developed the 100 kV EM 1932, whose resolution was 10 nm [3].

Metropolitan Vickers Company (Great Britain) started marketing the electrostatic-field type 1937 [4]. RCA (USA) started marketing 1940, Philips (Holland) also started marketing EM100 microscope 1946 [5], and AEG-Zeiss (Germany) developed the electrostatic-field EM-8 1949 [6]. Siemens (Germany) also developed the EM positively.

N. Yoshimura, *Historical Evolution Toward Achieving Ultrahigh Vacuum in JEOL Electron Microscopes*, SpringerBriefs in Applied Sciences and Technology, DOI 10.1007/978-4-431-54448-7_2, © The Author(s) 2014

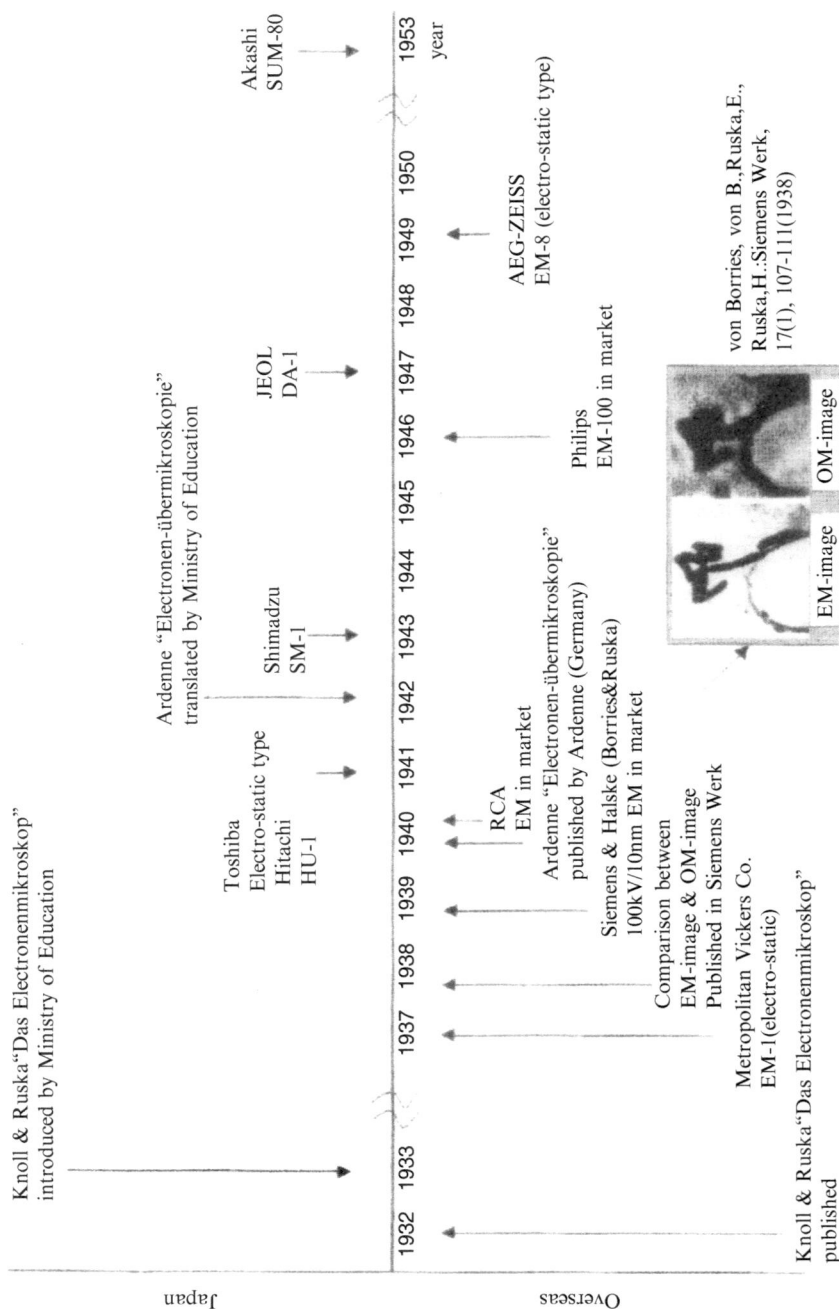

Fig. 2.1 History around EM in the world 1932–1952 [1]

On the other hand, several companies in Japan developed the EM several years later, as follows. Hitachi developed the HU-1 microscope (magnetic-field type) 1941 [7]. Shimadzu the SM-1 (magnetic-field type) 1943 [8], Electron Science Laboratory, predecessor of JEOL, the DA-1 (magnetic-field type) 1947 [9], respectively. Also, Toshiba developed the electrostatic-field type 1941 [10]. Lately, Akashi developed the SUM-80 (magnetic-field type) 1953 [11].

It is noted that most of the pioneers for EM were big companies manufacturing various electric instruments representing the respective nations, like as RCA for USA, Philips for Holland, Siemens for Germany, and also Hitachi and Toshiba for Japan. On the other hand, JEOL challenged developing EM, a little bit later, not having any other commercial instruments. It is also noted that many of such big companies withdrew from the commercial field of EM. Today, only JEOL Co. Ltd., is developing the commercial activity with analytical EMs (AEMs) from small-size EMs to ultrahigh-tension (UHT) EMs (UHT-EMs).

2.2 Dawn and Developing Periods (1946–1989) of JEOL Electron Microscopes

Let us see the course of history of JEOL electron microscopes, JEMs, from the history of our company, JEOL (Japan Electron Optics Laboratory) [12, 13].

Mr. Kenji Kazato, the leading person of the founders, called the young engineers who had worked together at the navy's laboratory to gather to develop the EM. Several members gathered in Mobara-cho (in Chiba prefecture). Dr. Kazuo Ito, the key person in developing the EM, also responded to the calling of Mr. K. Kazato [12].

2.2.1 Birth of AD-1 EM (1947)

The birth of AD-1 EM, assembled on a wood table, was the debut of our company, JEOL. It was September 1947, a few years after the end of the Pacific War [12].

There are two kinds of lens system, the electrostatic-field lens system and the magnetic-field lens system. At the beginning of developing EM in Germany, two kinds of lens systems were investigated side by side. The engineers at Mobara-cho had to decide first which lens system to be adopted. And, the electrostatic-field lens system was adopted by the reason that the required stability of high-voltage power supply for the magnetic-field lens system could not be expected at that time.

Dr. K. Ito described that the translation version of "Elektronen-Übermikroskopie" by Ardenne, M. v. [14] served well in theoretical designing.

The trial microscope was assembled, resulting in failure to get an EM image. Many reasons were there for failure, as follows. Vacuum in the column was not so good, the electrostatic-field lens system did not work well, and some of the electric parts applied could not work well, and so on.

The magnetic-field lens system was decided to be adopted at the meeting in company, instead of the electrostatic-field lens system, 10th of June, 1947. This decision
must be very important in developing EM successfully, considering that the preced-
ing companies (AEG in Germany, GE in USA, Toshiba in Japan), manufacturing
EMs with electrostatic-field lens systems, withdrew from the market.

The first JEOL EM with the magnetic-field lens system was designed and manufactured rapidly. The power supply with good stability was obtained by applying
a large capacitance to the power supply for X-ray instruments. And, at last, the first
EM with magnetic lens, DA1, presented in Fig. 2.2, was manufactured completely at
the end of September, 1947. The EM image of zinc oxide was observed by DA-1 2nd
of October, 1947. It was 1 year and half after starting investigation of EM in
Mobara-cho.

The first one of the DA-1 was sold 550,000 yen to Mitsubishi Chemical Co Ltd.
December 1947, which was used for investigating the surface construction of ion-
exchange resin. The DA-1 and DA-2 received orders from laboratories of govern-
ment, companies, hospitals and colleges, totally eight orders.

However, the actual situation of DA-1 must be recognized to be a trial microscope assembled on a wood table. The resolution of DA-1 was about 5 nm, which
should be compared with 2 nm of the microscope made in USA [12].

JEOL Co Ltd. was established formally 30th of May, 1949.

Fig. 2.3   JEM-1, by courtesy
of JEOL Co. Ltd

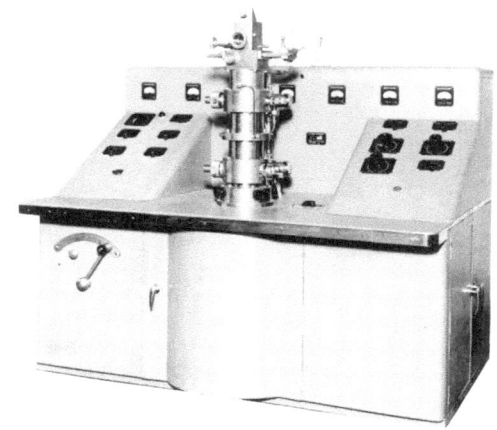

2.2.2   JEM-1 (1949)

The first one of JEM-1 with a magnetic-field lens system, presented in Fig. 2.3, had the following features.

- Upside-down column
- High-frequency power supply
- Equipped with an electron-diffraction imaging system
- Two-stage lens system without a condenser lens

JEM-1 had several defects. The big problem was that dust of the specimen fell to the anode aperture, making regular operation difficult. And, adjustment of brightness was impossible because JEM-1 did not equip with any condenser lens. Resultantly, JEM-2 and following ones returned to the regular column type.

JEM-## series continued up to JEM-7, which guaranteed the highest resolution at that time (1964). The idea that the adjustment of lens axis should be done electrically, was adopted in JEM-7 first. This idea was succeeded by the next microscope JEM-100B and successors, in which the electro-magnetic deflection unit and the electro-magnetic stigmator were installed.

2.2.3   JEM-100B (1968)

Figure 2.4 shows the cover of the catalog of JEM-100B, presenting the molecular image arranged in the copper chloride phthalocyanine crystal, which was firstly observed by JEM-100B. This pattern of the shell of a tortoise raised the evaluation of JEM-100B. Another factor to raise the evaluation was the followings. At 7th International Conference at Grenoble, France, August 1970, the EM image of single

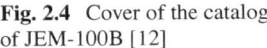

Fig. 2.4 Cover of the catalog
of JEM-100B [12]

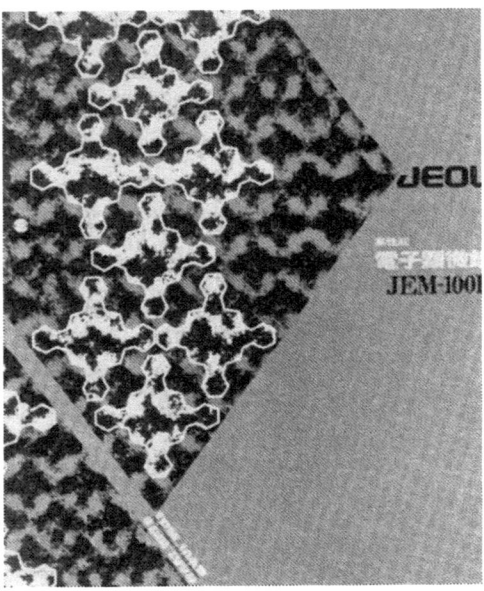

crystal of graphite on the florescent screen was demonstrated to be seen by the participants at the Conference. Usually, such an epoch-making photograph of the EM image was obtained by a skillful operator at midnight when the power voltage is stable and no vibration was expected, by making the electron beam be parallel and only one was selected among many photographs. It was good demonstration of JEM-100B to make the participants at the conference see such EM image on the fluorescent plate of the microscope, installed on the floor of the exhibition hall. That is, JEM 100B itself demonstrated how stable it is for vibration, power supply, and thermal conditions.

JEM-100B, making debut August 1968, was guaranteed 0.2 nm resolution. The eucentric drift-free specimen stage, installed in this microscope, was a unique unit to make the specimen drift 1/10, making dark selected-area image obtained with high resolution. Besides, other unique units such as the image wobbler and the top-entry specimen stage with six specimens, the digital display of magnification and the high-brightness electron gun, were installed.

JEM-100B first made the molecular-image observation possible. Both of specimen damage and contamination build-up were made sufficiently low by means of the new DP-evacuation system (see Sect. 4.1.3).

The patent (Patent 983332, 1970) made the combined microscope of TEM and SEM possible. This SEM-image observation unit was successfully developed 1970 and installed in JEM-100B. In this way, JEM-100B had the three microscope functions, the transmission EM (TEM), the scanning secondary EM (SEM), and the scanning transmission EM (STEM). Furthermore, JEM-100B was equipped with an EDS (energy-dispersive X-ray spectroscope). That is, JEM-100B made its debut first as an analytical EM (AEM) in the world [12].

2.2.4 Analytical Electron Microscopes with a Cascade DP System (After 1973)

JEM-100C and many successors of analytical electron microscope (AEMs), 100CX/200CX/1200EX/2000EX…, equipped with the cascade DP evacuation system. The cascade system was refined year by year, which is described in details in Sect. 4.1.4.

In the specimen chamber of JEM-100CX, the vacuum pressure of 10^{-5} Pa range and the resolution of 0.14 nm without using a liquid-nitrogen trap were obtained. Furthermore, a LaB_6 emitter could be applied to the electron gun, which served well by emitting high-current-density electron beam. JEM-100C/CX series was indeed best sold and are still used by many researchers. More than one thousand units of JEM-100C/CX series had sold by February 1984, which was indeed the surprised record as the high-price AEM [12].

The successor of JEM-100C/CX series, JEM-1200EX was developed March 1978, which was provided with the refined cascade DP system. JEM-1200EX had the image-forming lens system composed of two-stage objective lens, three-stage intermediate lens and one-stage projection lens, which made rotation-free EM-image observation possible from multiplication of 50× to 1,000,000×. An SIP (sputter ion pump) and a TMP (turbo-molecular pump) were optionally provided, if desired by customers.

2.2.5 High-Tension Electron Microscopes

JEM-500D, 500 kV cryogenic EM of ultrahigh resolution was delivered to Institute for Chemical Research, Kyoto University, March 1974. This microscope was the product of the corporation of JEOL and Profs. K. Kobayashi and N. Ueda of Institute for Chemical Research. The microscope was developed in order to observe the atomic and molecular construction directly in the state of the specimen being held at cryogenic temperature (4.2–7 K). JEM-500D, cryogenic 500 kV EM achieved the resolution of 0.14 nm even at 500 kV. This high resolution was achieved through many improvements, as follows: Two high-tension (HT) tanks were provided for installing the HT generator and the HT electron gun, separately. Multiple feedback circuits were assembled in order to make the stability of the acceleration voltage be extremely high. The water-cooled drift-free specimen stage was provided. And, in order to minimize the contamination build-up, the cryogenic trap, cooled by liquid helium, was installed both in the mini-lab chamber and just under objective lens. Through the development mentioned above, the photograph of the atom-arrangement of vanadium penta-oxide was obtained by JEM-500D first in the world.

Increasing the acceleration voltage makes the penetrating power of electron beam be increased. That is, using 200 kV electron beam the 1.6× thicker specimen can be observed at the same brightness, comparing with 100 kV electron beam. At the same time, the tolerable limit of the electron intensity for the specimen being damaged due to irradiation also become enlarged as the transmission power enlarged. That is, HT electron-beam observation is suited to the biological specimen to be damaged easily. Furthermore, thanks to their very small chromatic aberration of 2.6 mm, the resolution of the JEM-200CX series was more than two times that of 100 kV electron microscopes. And, the JEM-200CX series guarantied the high resolution of 0.14 nm.

Needs for the researchers on the biological field to use the HT-EM became larger and larger.

The first one of the 1,000 kV EM was manufactured March of 1966 and the fruits of which were presented at the International Conference of Electron Microscopy, held August of 1966. The ultrahigh-tension (UHT) EM was further improved on the acceleration tube and high-voltage generator circuit, and resultantly routine operation of the 1,000 kV EM became possible 1967.

The needs for UHT-EM were conventionally to research the materials for atomic power science. On the other hands, new needs in the semiconductor field became bigger and bigger, as the 200 kV EM got the high evaluation for atomic-resolution observation of the lattice defect of ICs. However, the price of the atomic-resolution 1,000 kV EM (UHT-ARM) was really high as nearly twenty times of the price of JEM-200CX, and the new special building for the UHT-EM was additionally necessary, making general institutes buying the UHT-EM difficult. In order to resolve this price problem a 400 kV EM was being developed, parallel for 1,000 kV microscope. The first 400 kV EM was installed in the Department of Technology, Osaka University, which was co-developed for atom-resolution 400 kV-EM with JEOL, March 1983. January of 1984 JEM-4000EX was commercially completed.

JEM-4000EX guarantees 0.19 nm of point resolution and makes it possible to micro-area analysis by adopting three-stage condenser-lens system. JEM-4000EX can be installed in the conventional building because of its compact size (3.1 m height, 4 t weigh) [12].

2.3   Nearest 20 Years (1989–2010) in the History of JEMs

JEOL considered this period as "developing and maturity period of JEMs".

2.3.1   JEMs with a Field-Emission Electron Source

It was natural that a fine, high-current density electron source was desired for EMs. The LaB$_6$ thermal emitter was used instead of the W-thermal emitter, and the field-emission (FE) sources (cold type and thermal one) were desired.

JEM-2010, which was put place on the market 1989, equipped with the newly developed specimen stage for inserting the specimen into the pole-piece gap of the objective lens. And JEM-1210 (120 kV), -3010 (300 kV), -4010 (400 kV) and -2010 F (200 kV with FE emitter) were lined-up, all of which equipped with the new specimen stage mentioned above.

Prof. Browning of Illinois University observed the STEM image of 0.134 nm-dumbbell of Si<110> single crystal first in the world by using JEM-2010F/STEM with FE emitter, which led the high-resolution STEM to be received the high evaluation. Furthermore, the Cs-corrector was installed to JEM-2200FS and -2100F 2003, which made the atomic-resolution analysis possible [13].

2.3.2 Ultrahigh-Tension EMs (UHT-EMs)

JEM-ARM1000, delivered for Institute for Chemical Research, Kyoto University 1989, was provided with the twin-tank UHT generation system in order to ensure the improvement of voltage stability by separating the voltage generation circuit and the acceleration tube, respectively.

On manufacturing JEM-ARM1250 for MPI (Max-Planck-Institute) of Germany, the building, provided with anti-vibration supporting system, was constructed in JEOL Akishima factory in order to check its guaranteed performance before delivery. After delivery of JEM-ARM1250 1993, a new circuit was provided to improve the stability of UHT for achieving the theoretical resolution with a type of the top-entry goniometer specimen stage. And finally, the world-recorded point resolution of 0.1 nm was achieved.

The UHT-EM delivered to NRIM (National Research Institute for Metals, Japan) 1994 and one to Center for Advanced Research of Energy and Materials, Hokkaido University, Japan 1998, were UHT-EMs with the side-entry goniometer (SEG) specimen stage, equipped with the large ion-thinning device for in situ observation. By using these UHT microscopes with SEG specimen stage, EM images of the theoretical resolution were observed, which made the UHT microscope with SEG specimen stage get high popularity.

The UHT-EM delivered to KBSI (Korea Basic Science Institute) 2003 had the function of remote control by which many users can utilize the microscope with keeping high resolution [13].

JEM-1300NES, delivered to Kyushu University 2005, equipped with the Ω-filter, which served well for thick specimens to observe the high-contrast bright image [13].

JEM-ARM1300 (atomic resolution, 1,300 kV), presented in Fig. 2.5, permits the direct observation of 3-D structure of thick specimens and direct observation of sensitive biological specimens with little damage. It features great operational ease and stability [13–15]. The twin-tank UHT generation system of JEM-ARM1300 is presented in Fig. 2.6 [13].

2.3.3 Cryogenic EM

The helium (He)-stage received high evaluation after the superfluidity He-stage was developed 1986 with co-operation of Prof. Y. Fujiyoshi and JEOL. Prof. Y. Fujiyoshi and JEOL took notice that when the specimen was kept at cryogenic temperature the damage the specimen received when being illuminated with an electron beam

Fig. 2.5   JEM-ARM1300, by courtesy of JEOL Co. Ltd [16]

was much reduced, and developed the superfluidity He-stage for construction analysis of biological specimen. The performance of this cryo-stage was superior to the stage in general use, as follows. Cryo-temperature; 4.2 K, lattice resolution; 0.2 nm. This cryo-microscope, provided with the cryo-transfer mechanism for cryo-specimen, serves well for construction analysis of protein [13].

2.3.4   JEMs with a Cs-Corrector

The topic in 2000s was the Cs-corrector.

CEOS company of Germany put the Cs-corrector for EM in market, which made JEOL contact with CEOS for the Cs-corrector. And, JEM-2010F with the Cs-corrector finally achieved the expected resolution of 0.1 nm [16]. According to the expected effect of the Cs-corrector, JEOL actively installed the Cs-corrector into

Fig. 2.6 The twin-tank UHT generation system of JEM-ARM1300, by courtesy of JEOL Co. Ltd. [13]

the 200 kV-EMs. That is, JEM-2100F with the Cs-corrector was delivered to Nagoya University (Prof. Tanaka) 2003 [17], the Cs-corrector for STEM was delivered to Tokyo University (Prof. Isohara) in order to be installed into the JEM-2100F under usage [18]. And the JEM-2200F with the Cs-corrector for TEM/STEM was delivered to Oxford University 2003 [19].

A new atomic-resolution microscope, JEM-ARM200F (200 kV, FEG) was on the market 2010, that is, 60 years after JEOL birth.

The JEM-ARM200F, incorporating a STEM Cs corrector and a microscope column with dramatically improved mechanical and electrical stability, achieves the world's highest STEM resolution of 0.08 nm. The Cs-corrected extremely small electron probe achieves a remarkably increased current density, one order magnitude larger than conventional TEMs. Thus, the JEM-ARM200F provides ultimate atomic-level analysis and also higher throughput with dramatically shortened measurement time [15].

The JEM-ARM200F, as one of the latest JEMs, represents the flagship of JEOL, from its birth.

2.3.5   Prospect of the Resolution of EMs

Figure 2.7 shows the transition of resolution of TEM/STEM, from the invention of EM in 1932 to the present. Today, the resolution of 0.1 nm can be obtained by the help of the C_S corrector and the FE electron sources, together with the help of the ultrahigh-vacuum technology and anti-vibration technology for microscope column.

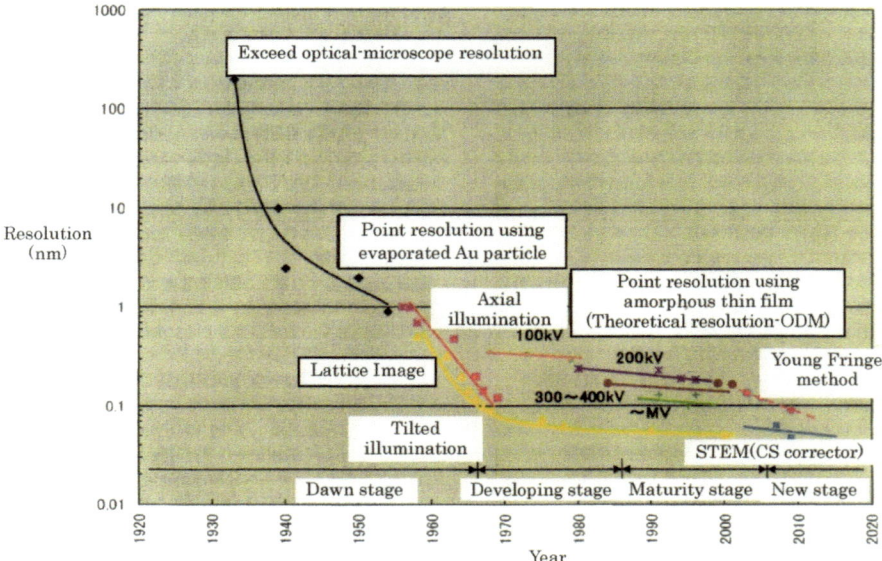

Fig. 2.7 Transition of resolution of TEM/STEM [1]

In order to develop the TEM/STEM with further high resolution, not only the further development of electron sources and anti-vibration technology is required, but also the development of application field is more important. We believe the wider field of science in near future will require the TEM/STEM/AEM with extremely high resolution.

References

1. Harada Y, Tomita M (2011) Progress of electron-microscope technology in Japan. Microscopy 46(Suppl 3):1–47 (in Japanese)
2. Knoll M, Ruska E (1932) Z Physik 78:318–339
3. Von Borries VB, Ruska E (1939) Naturwiss 27:577–582
4. Martin IC, Whelpton RV, Parnum DH (1937) J Sci Instrum 14:14–24
5. van Dorsten AC, Nieuwdorp H, Verhoeff A (1950) Philips Tech Rev 12(2):34–51
6. Fujita H (ed) (1986) History of electron microscopes. Published in commemoration of 11th ICEM, Kyoto, p 212
7. Komoda T (1986) History of electron microscopes. In: Fujita H (ed) Published in commemoration of 11th ICEM, Kyoto, pp 49–53
8. Shimadzu S (1986) History of electron microscopes. In: Fujita H (ed) Published in commemoration of 11th ICEM, Kyoto, pp 58–63
9. Ito K (1986) History of electron microscopes. In: Fujita H (ed) Published in commemoration of 11th ICEM, Kyoto, pp 54–57
10. Kamogawa H (1986) History of electron microscopes. In: Fujita H (ed) Published in commemoration of 11th ICEM, Kyoto, pp 64–79

11. Fujita H (ed) (1986) History of electron microscopes. Published in commemoration of 11th ICEM, Kyoto, p 85
12. JEOL Co. Ltd. (1986) History of JEOL (1); 35 years from its birth, March 1986 (in Japanese)
13. JEOL Co. Ltd. (2010) History of JEOL (2); creation and development: 60 years from its birth, May 2010 (in Japanese)
14. Ardenne MV (1942) Electron microscope by Ardenne MV, Springer, Berlin (translated into Japanese from Ardenne MV). Ministry of Education, Maruzen
15. Product guide (2012) No. 0403C278C Printed in Japan, JEOL Co. Ltd.
16. Hosokawa F, Tomita T, Naruse M, Honda T, Hartel P, Haider M (2003) J Electron Microsc 52(1):3–10
17. Tanaka N, Yamazaki J, Saito A (2008) J Vac Soc Jpn 51(11):695–699 (in Japanese)
18. Isohara Y (2008) J Vac Soc Jpn 51(11):700–706 (in Japanese)
19. Sawada H, Tomita T, Naruse M, Honda T, Hambridge P, Hartel P, Haider M, Hetherington C, Doole R, Kirkland A, Hutchison J, Tichmarsch J, Cockayne D (2005) J Electron Microsc 54(2):119–121

Chapter 3
Accidents and Information, Instructing Us to Improve the Vacuum Systems of JEMs

Abstract Some accidents sometimes occurred on the vacuum systems of JEMs. We must resolve them as soon as possible. A backstreaming accident occurred on a customer's EM was, indeed, a most severe one for us, the vacuum engineers of JEOL. In that case the lead-glass window of the viewing chamber got gloomy after EM-film magazine was exchanged.

Accidents and information on the vacuum systems of the JEMs always instructed us how to improve the vacuum systems of JEMs.

3.1 Accidents Occurred in the Vacuum Systems of JEMs

Accidents, occurred on the vacuum systems, always instructed us to improve the vacuum systems of JEMs.

3.1.1 Porosity of the Wall of Camera Chamber, Made of Brass Casting

The viewing chamber and the camera chamber were made of brass casting or steel casting. The reasons are as follows.

- Manufacturing such chambers of complex shape by casting is very cheap in cost compared with manufacturing by machining and welding.
- Brass or steel is suited for shielding the X-ray emitted from the metal walls, irradiated with a high-speed electron beam.
- Large weight of the viewing chamber is desired as the base for assembling the microscope column. In order to minimize the vibration of the column, the larger the weight of base-plate, the better.

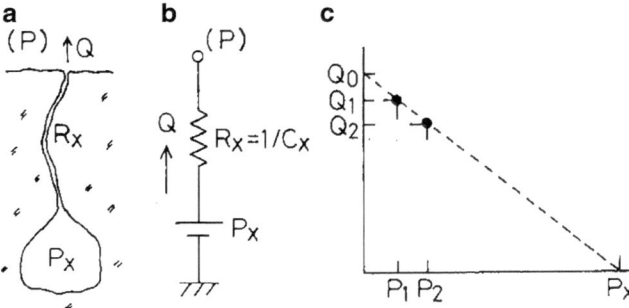

Fig. 3.1 Concept of outgas source: (a) gas reservoir and capillary, (b) pressure generator with Px and Rx, and (c) characteristic values, Px and Q_0 [1]

On the other hand, the porosity of casting material is well known as a large source of outgassing.

Information from the manufacturing department was as follows. "The color of discharge of the Geissler's tube attached to the camera chamber is keeping red in spite of a long-time evacuation. Leek test by spraying ether solvent shows no air-leak."

We knew that leak checking for the camera chamber is quite difficult because of the narrow capillary of pores in casting walls. Engineers in the manufacturing department were well aware how to check leakage of casting material. We suspected that many deep cavities with long, narrow capillaries existed in the wall of the camera chamber. We decided to check the said camera chamber thoroughly. "Let's evacuate the said camera chamber for a long time until the color of discharge of the Geissler's tube will turn to blue or white, showing that almost all air in all cavities has been pumped out."

It was the next morning about 20 h after starting evacuation of said camera chamber by a mechanical rotary pump (RP) that the discharge color turned to the mixed color of white and blue, showing air in all cavities had been pumped out. The said camera chamber was judged to be usable as a test-passed chamber.

The basic concept of the porosity of casting wall led me the idea of simulating the high vacuum system by using a resistor network. Please watch Fig. 3.1 carefully and find out the meaning involved. You will understand that both of the functions of an outgassing source and a high-vacuum pump are same in essence and that they can be represented by the pressure generator connected to ground level of zero pressure. When the characteristic pressure Px is lower than the pressure P in the high-vacuum chamber, gas molecules in the chamber flow into the vacuum generator. That is, when Px > P, the pressure generator functions as an outgassing source, and when Px < P, the pressure generator as a high-vacuum pump. This idea led me establish the resistor network simulation method to get the pressures and gas flows at every positions [1]. (See Sect. 4.2.2 for the resistor network simulation method.)

3.1.2 Viton O-Rings, Softened and Swollen when Used in Freon-Gas Environment

Several Viton O-rings of a large size were sent us from the service department, which were used in the Freon-gas tank for sealing the multi-stage accelerating tube. We were surprised to see that these O-rings were softened and swollen. It was apparent that these O-rings could not be used for sealing.

Viton O-ring is made of fluorine elastomer, which is chemically same as that of Freon-gas. We considered that the said O-rings were softened and swollen when being used in Freon-gas environment for a long period. Freon gas must leak out into the accelerating tube, causing a trouble of vacuum.

The accelerating tube of the UHT EM of the customer was, of course, thoroughly cleaned and new Viton O-rings were applied for sealing. And, we must resolve this problem.

Viton O-rings are suited for high vacuum because of their low outgassing rates. On the other hand, Nitril O-rings are not swollen when used in Freon-gas environment, though their outgassing rates are rather large. The idea to resolve this problem was to apply the double O-rings seal. Nitril O-rings are applied to the Freon-gas side and Viton O-rings to the vacuum side.

This idea of double O-rings for sealing of accelerating tube in Freon-gas tank was registered as a patent [2].

3.1.3 An Extremely Large Amount of Water Vapor, Evolved from a Specimen Stage of Anodized Aluminum

Information from the adjustment department was as follows.

"The carbon-film specimen is etched when illuminated with an electron beam, losing its mass." The fins of ACD (anti-contamination device) of the said microscope were cooled by copper mesh wires being soldered to a copper tank containing liq. N_2. The said microscope equipped a specimen stage made of anodized alumina by way of trial.

First, we considered as follows: "ACD has a high pumping speed for water vapor, so when the fins of ACD are cooled with liq. N_2, partial pressures of H_2O must be reduced to a usual level. Therefore, an air-leak must exist in the vicinity of the specimen." "Let's hunt the leak location using a mass-filter."

The residual gas pattern when the ACD fins were at room temperature showed that the peak of M/e = 18 (H_2O^+) was very high and the peaks of M/e = 32 (O_2^+) and M/e = 28 (N_2^+, CO^+) were rather low, showing a leak-free pattern. And, the residual gas pattern when the ACD fins were cooled with liq. N_2 also showed that the peak of M/e = 18 (H_2O^+) was still very high and the peaks of M/e = 32 (O_2^+) and M/e = 28 (N_2^+ and CO^+) were still low, showing a leak-free pattern. That is, abnormally high H_2O pressure was considered as the cause of specimen etching.

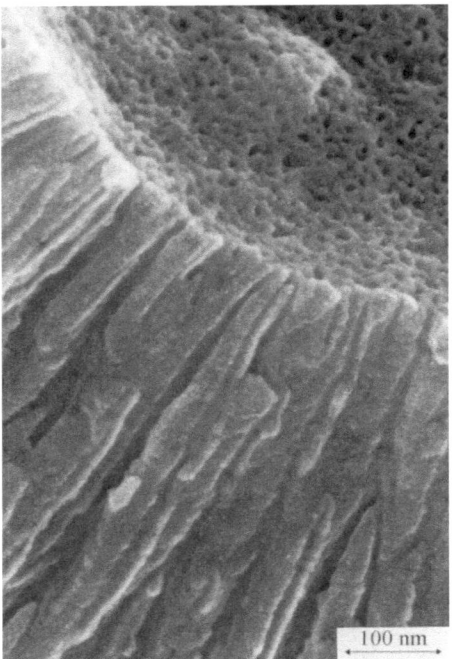

Fig. 3.2 SEM image of anodic oxide layer of aluminum, by courtesy of JEOL Ltd

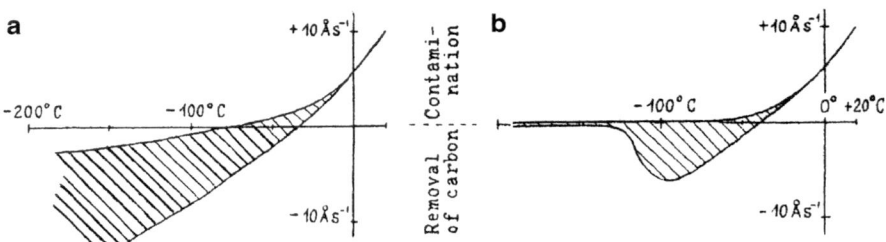

Fig. 3.3 Rates of contamination build-up and carbon removal as functions of the cool chamber temperature. Beam-current density at object 0.4 A/cm², diameter of irradiated area about 1.8 μm, total pressure in microscope 2 to 5×10^{-5} Torr. (**a**) Specimen temperature ≅ Cool chamber temperature. (**b**) Specimen temperature ≅ Room temperature. (Heide, 1962) [3]. *Note*: 1 Torr (or torr) = 133.3 Pa

We knew the abnormal porosity of the wall of anodized aluminum, by seeing its SEM-image in Fig. 3.2. Abnormally large amount of H_2O molecules must be evolved from the anodized surface with narrow capillaries and deep vacancies.

I had read an important paper titled "The preventionof contamination without beam damage to the specimen" by Heide [3]. The paper reports the experimental results that when the carbon specimen is illuminated with an electron beam in the vacuum environment with high H_2O pressure and low hydrocarbon pressure,

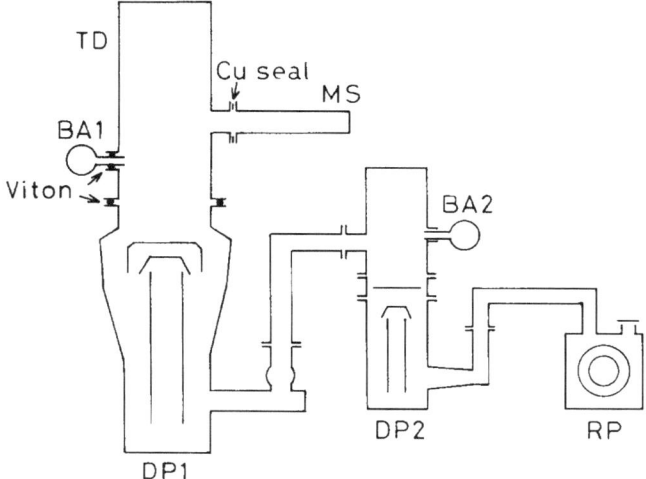

Fig. 3.4 Experimental setup [4] A thick Viton O ring (Co-seal, 4 in.) is used between the test dome (TD) and the first diffusion pump (DP1, Diffstak 100/300)

the specimen is easily etched. His experimental result is presented in Fig. 3.3. The paper also presents the information that in the electron microscope this phenomenon easily occurs when the temperature of the cool trap surface is around −100 °C. The conditions of our case is almost the same as those of Heide [3].

The problem of etching was resolved by inhibiting the trial stage made of anodized alumina in the microscope column.

3.1.4 H_2O Molecules in the Atmosphere, Permeating Through Viton O-Ring Seals

The residual gas spectrum analyzed in the specimen chamber shows H_2O^+ peak is usually the highest one. It is a natural question what is the source of water vapor.

A DP evacuation system was running for a long time after the test chamber had been baked at about 100 °C for about 3 days to ascertain the clean vacuum to be kept. We happened to discover the source of water vapor when watching the residual gas spectrum, as mentioned below:

The experimental setup is shown in Fig. 3.4. A stainless-steel test dome (TD) was evacuated by the first pump [DP1, Diffstak 100/300, 280 L/s, Santovac-5 (polyphenylether), Edwards] followed by the second one (DP2, 2.5 in., Santovac-5, JEOL). A thick Viton O-ring (Co-seal, 4 in., Edwards) was fitted between TD and DP1. BA1 was sealed with a small Viton-A O-ring, P-14 (Mitsubishi Cable Industries). MS was sealed with a copper gasket. The experiment was performed in summer for about one month. The room temperature and humidity during the period were about 25 °C and 75 % on an average, respectively.

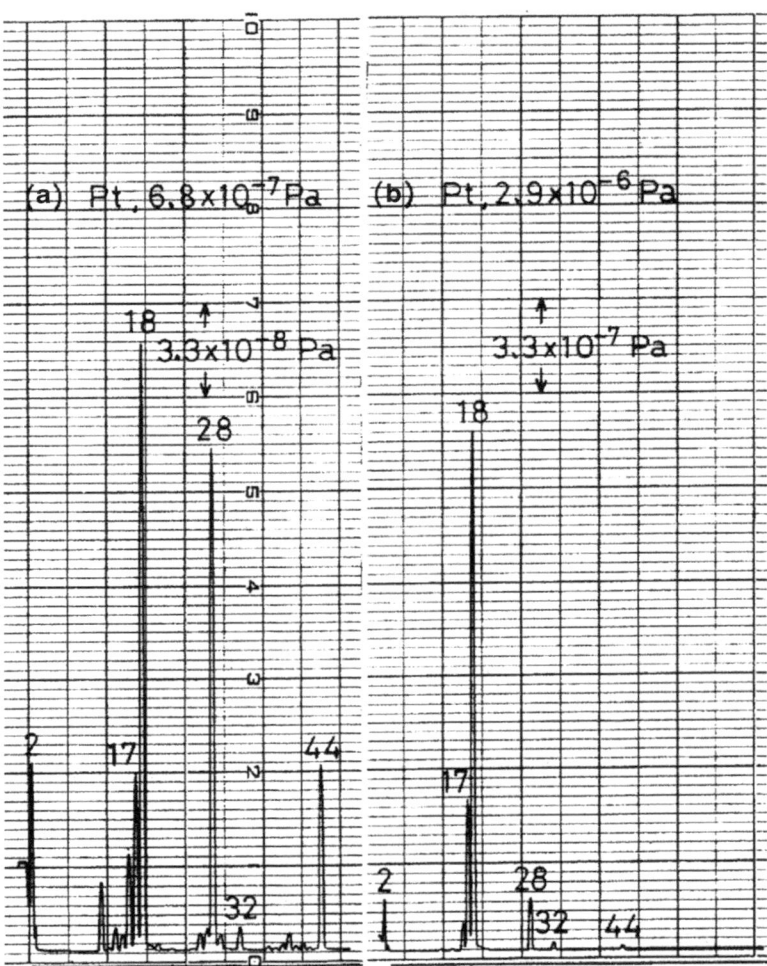

Fig. 3.5 Residual gas spectra: (**a**) on the second day after baking and (**b**) on the 29th day [4]

The TD with MS was first baked (about 100 °C, for 3 days) in situ under high vacuum. The bakeout temperature of the Co-seal was guessed to be nearly 50 °C. Mass spectra on the 2nd day and 29th day after baking are shown in (a) and (b) of Fig. 3.5, respectively. Hydrocarbon peaks decreased in first few days, and were kept in minimum levels. On the other hand, H_2O peaks [masses 17 (OH^+), 18 (H_2O^+), and 19 (H_3O^+)] were kept almost constant in the first several days, and then increased gradually until saturated. O_2^+ peak (mass 32) also increased with pumping time until saturated. H_2^+ peak (2) increased slightly with time. Other atmospheric gases, CO and N_2 (28), Ar (40), and CO_2 (44), on the other hand, did not increase with time. The rise of total pressure (P_t), H_2O pressure (17, 18, 19), and O_2 pressure (32) are presented in Fig. 3.6.

Fig. 3.6 Pressure-rises with pumping time. Total pressure P_t and pressures of masses 17 and 18 are read by the left-hand ordinate, and pressures of 19 and 32 by the right-hand ordinate [4]

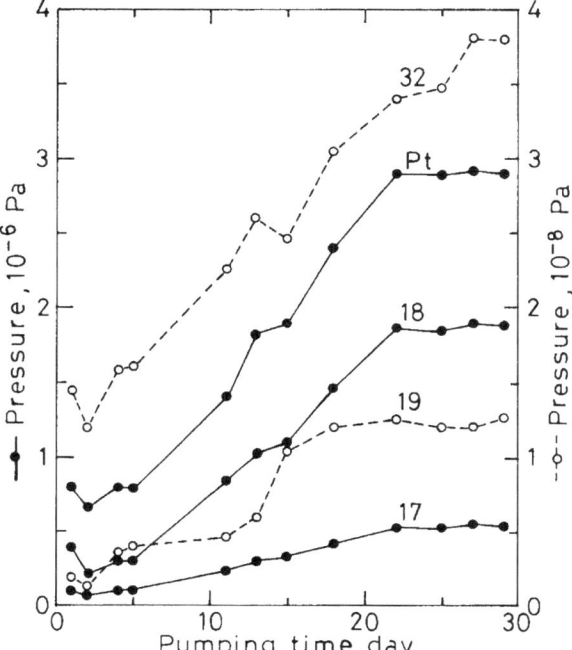

A modified setup was provided, in which Diffstak 100/300 of DP1 in Fig. 3.4 was replaced by the stack of another DP with a cold cap and a water-cooled chevron baffle (4 in., 170 L/s, Santovac-5, JEOL). As a result, two Viton O-rings, G-120 and G-110 (Mitsubishi Cable Industries) were fit on both faces of the baffle.

Experiment was performed on the modified setup in winter for about one month. The room temperature and humidity during the period were ~20 °C and ~40 % on an average, respectively. TD with MS was again baked (~100 °C, 3 days) under high vacuum. The bakeout temperatures for Viton O-rings G-120 and G-110 were guessed to be 40–50 °C.

The mass spectrum on the 30th day is shown in Fig. 3.7. The sum of H_2O peaks (17, 18, 19) represents a concentration of about 75 %. The spectrum shows a very clean vacuum with respect to hydrocarbon gases. Peaks of 14 (N_2^{++}) and 40 (Ar^+) are negligibly low, showing no air leak.

It is notable that peaks 17 and 18 are extremely high and that peak 32 is relatively high. The peak 32 is caused by the O_2 permeation through Viton seals.

It can be concluded that H_2O gas in the conventional high-vacuum system with Viton O-ring seals is due to the permeation through Viton O-rings.

Comments: The rate of permeation through Viton O-rings might depend on the Viton O-ring itself (makers of O-rings), the extent of compression of O-ring, depending on the shape and size of O-ring groove, temperature and humidity of the atmosphere. It should also be noted that oxygen gas (mass 32) permeates the Viton O-rings, as shown in the residual gas spectrum of Fig. 3.7.

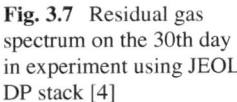

Fig. 3.7 Residual gas spectrum on the 30th day in experiment using JEOL DP stack [4]

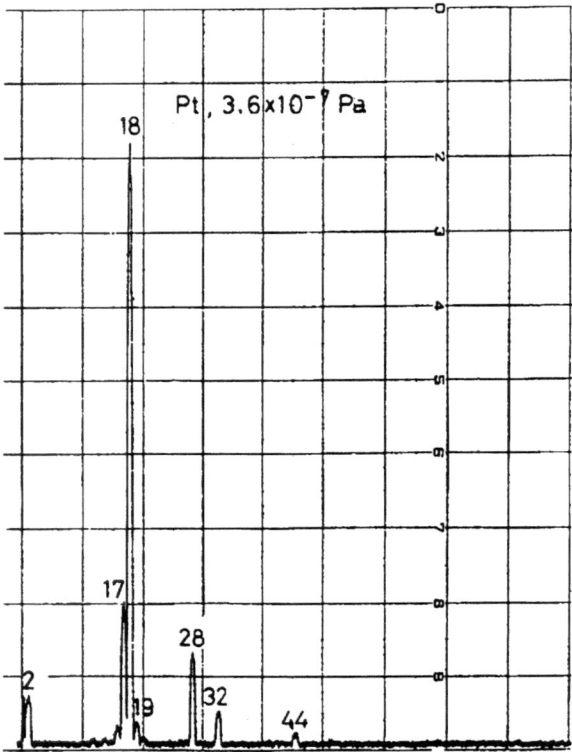

3.1.5 Jet-Nozzle Assembly of Diffusion Pump, Melted Thermally

A 4-in. JEOL diffusion pump (DP) was sent us from the service department of JEOL-USA. The jet assembly of aluminum alloy was melted thermally. The interior of the DP turned brown in color and DP fluid turned to be solid, being adhered to the inner walls of the pump vessel.

The said accident occurred in the following way. The customer applied the automatic liq. N_2 supply system to the customer's EM, resulting in the liq. N_2 trap being cooled for a long period. After using the automatic liq. N_2 supply system for several months, the system was switched off, with the EM being in the "stand-by" mode. (In "stand-by" mode, the evacuation valves for the whole column are closed, keeping the DP evacuation system be activated.) As the time passing, the temperature of thick ice on the surface of liq. N_2 trap rose gradually after the liq. N_2 tank got empty, and a large amount of H_2O vapor must be evolved with the DP heater being on. The pressure inside the DP vessel rose gradually much beyond the critical backing pressure during which the backing rotary pump was evacuating the DP vessel with the DP heater being on. As a result, the DP jet assembly of aluminum alloy melted thermally.

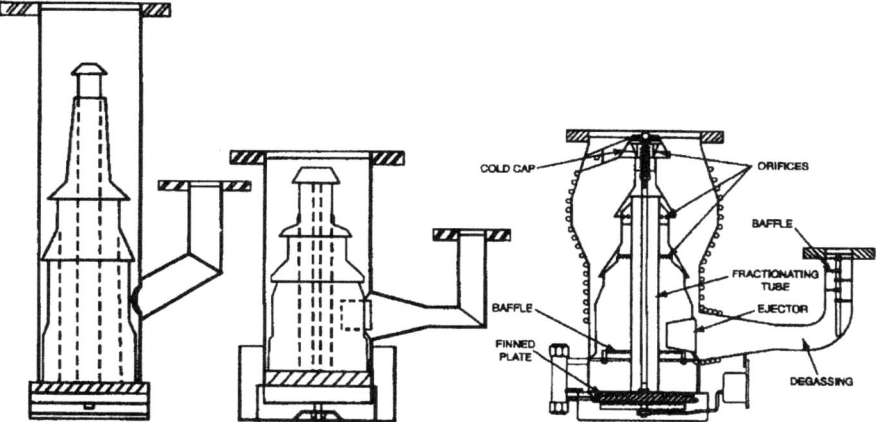

Fig. 3.8 Illustration of the progress in pump design: *left*—prior to 1958; *center*—1961; *right*—1965 [5]

The said accident instructed us to provide a safety system with the vacuum system. We discussed how to use a liq. N_2 trap, and reached the idea that the temperature of the cold surface of a trap must be raised to room temperature at least once a week with the EM being switched off.

Refer the cold trap presented in Fig. 4.13, which is suited to realize the idea that the temperature of the cold surface of a trap must be raised to room temperature at week end.

3.1.6   A Modern DP, Being a Vibration Source

An accident of vibration occurred on JEOL EPMA (electron-probe microanalyzer) being tested at the adjustment department. The said EPMA equipped with a modern DP (Varian) for customer's claim. We knew the features of Varian-DP from the paper by Hablanian and Maliakal [5], having high pumping speed and high throughput of gas.

The said DP is illustrated at the right side of Fig. 3.8. The pump has a thin-wall vessel of barrel shape and a long, wide diffuser and high heater power. The said DP was emitting a sound noise, which could be easily heard.

Notes

1. In the illustrations of the left-side and the middle ones in Fig. 3.8 water cooling pipes are omitted to draw. The water-cooling pipes were actually wound around the pump vessel and the ejector pipe for the middle pump, same as shown in the illustration of the right-side one. The water-cooling pipe was also wound around the pump vessel for the left-side one.

2. JEOL DPs in 1970s were almost same as one shown in the middle illustration, with the baffle in the fore-line pipe.

We suspected that the said modern DP was the source of vibration causing interference for EPMA analysis.

In order to suppress the vibration of the said DP, we inserted thin plates of lead tightly into the space between the cooling pipes soldered to the pump vessel by using a hammer. Making the mass of the wall of the pump vessel be increased showed a remarkable effect to suppress the vibration, resulting in removing interference for EPMA analysis.

How to Cut Off the Vibration from RP

The RP is set on the base plate which is isolated by rubber mount from the floor.

The evacuation pipes from RP and column are connected together using soft rubber tubes via a heavy concrete block. That is, the evacuation pipe of RP is first connected with a soft rubber tube to the pipe which penetrates the concrete block, and the evacuation port of the column is also connected with another soft rubber tube to another-side pipe from the concrete block. The vibration from RP is absorbed by the concrete block of large mass.

Sound from RP becomes a vibration source.

Heavy Viewing Chamber, Minimizing the Column Vibration

The metal base table, on which the microscope column is assembled, is isolated by metal springs from the floor.

The heavy viewing chamber is first installed on the base table, and the column is installed on the viewing chamber. The heavy viewing chamber absorbs the vibration from the RP and the floor.

3.1.7 Microdischarge, Occurred on Customer's JEM

Microdischarge occurred on the EM of the customer who observed many convergent-beam diffraction (CBD) images of crystallization materials. The claim was as follows. "When observing the CBD images, microdischsrge often occurs. The CBD images are affected by microdischarge, resulting in noisy ones."

We had been investigating microdischarge on the JEOL 100 kV-electron gun. Microdischarge occurred when the electron beam had been emitting for several hours continuously.

I got the information that the said customer often illuminated the specimen with an electron beam continuously for more than one hour. The situation of a long-time continuous illumination for the said microscope was quite similar to the situation when microdischarge easily occurred. Diffraction-image observation usually necessitates a long-time continuous illumination of an electron beam, for instance, several 10 min.

I visited the customer's laboratory to check the situation. The professor used to illuminate the specimen for more than one hour continuously, and after a short stop of illumination, another area of the specimen was illuminated continuously for a long time. I told him, "let us do an experiment on this microscope. I think microdischarge will occur in several hours."

The said microscope was ready to use at that time. I connected a pen recorder to the check terminal of acceleration voltage in order to check the stability of the electron beam during the period of illuminating a specimen with an electron beam. And, as was expected, voltage spikes appeared on the recorder chart, which showed occurring microdischarge.

I had similar data of microdischarge on a JEOL 100 kV electron gun, which was obtained by us. I said "let me explain the characteristics of microdischarge on the electron gun using our experiment data."

The characteristics of microdischarge of our 100 kV electron gun was submitted as a paper a few years later (1987) [6]. Experimental conditions and results were as follows:

Microdischarge depends on the concentration of electric field in the gun chamber. Equipotentials around and on the electron gun, simulated by the finite-element method, show that the electric field is relatively high near the top and side of the Wehnelt electrode. The holes of the Wehnelt and anode electrodes through which an electron beam passes are about 0.4 and 6 mm in diameter, respectively. The surfaces of the electrodes and chamber walls (SS304) are mirror-polished. The insulator is a kind of porcelain whose surface has been treated to be glassy. The junction of metal and insulator is covered by a guard ring (SS304, mirror polished) in electrical and mechanical contact with the Wehnelt electrode.

Experiments were made using a typical DP evacuation system. -100 kV is applied to the Wehnelt electrode, with the anode electrode being kept at ground potential.

Microdischarges occurred most frequently on C-1 (1st day after CH_3CCl_3 cleaning) without AGC (Ar-glow conditioning), as presented in Fig. 3.9, where microdischarges are identified as voltage spikes on an oscilloscope and on a recorder line. Microdischarges began to occur at about 2 h and occurred very frequently in the period from 2 to 3.5 h after starting the illumination.

We considered that microdischarges depending on the elapsed time must be related to outgassing of the electron gun heated by the tungsten-filament emitter. During electron beam extraction, the heat (about 10 W) from the filament emitter must gradually raise the temperature of the insulator, resulting in increased outgassing from the insulator surface.

Outgassing from the insulator was examined by measuring pressure-rise caused by the lighted filament emitter and by analyzing evolved gases, in the chamber evacuated an 160 L/s sputter ion pump (SIP). Pressure variations due to outgassing from the heated insulator are presented in Fig. 3.10.

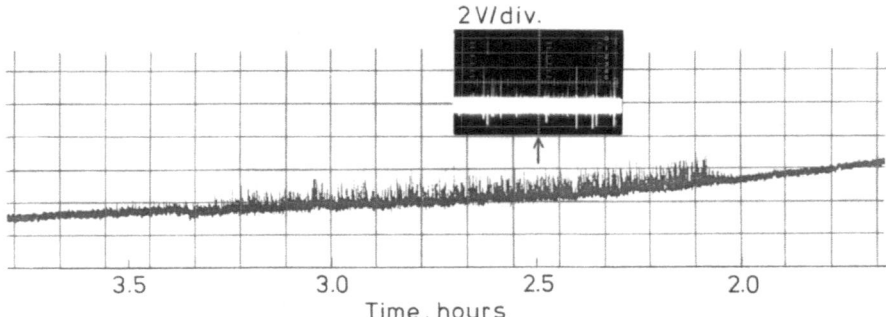

Fig. 3.9 Microdischarges under about 10^{-4} Pa on "C-1" (1st day after cleaning) without AGC (Ar-glow conditioning) near the beginning of the elapsed time. Microdischarges can be identified as voltage spikes on the chart line and on a photograph taken from the oscilloscope [6]

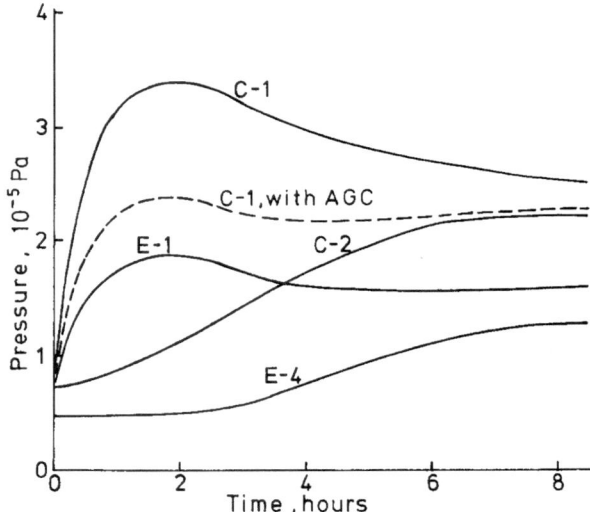

Fig. 3.10 Pressure-changes in the SIP system after the insulator being heated by the filament of about 10 W. The system was continuously evacuated by the SIP (160 L/s). C-1: 1st day after cleaning without AGC (Ar-glow discharge treatment); E-1: 1st day after exposing the chamber to the atmosphere without AGC [6]

The microdischarge characteristics depending on the elapsed time, shown in Fig. 3.9, are indeed analogous to the outgassing characteristics from the insulator depending on the elapsed time after switching "ON" to light the filament emitter on C-1 presented in Fig. 3.10.

Microdischarges are enhanced by the outgassing from the insulator. The reason is that the insulator surface with a high density of gas molecules causes high-yield secondary-electron emission, leading to positive charging on the insulator surface.

Thermal degassing for the electrodes and insulators has a conditioning effect to reduce microdischarge.

I explained the characteristics of microdischarge on the JEOL electron gun. "When the electron beam is continuously extracted for more than two hours, microdischarge will occur due to the outgassing from the porcelain insulator heated by the thermal W-emitter." The professor understood my explanation why microdischarge occurs when the electron emission was being "ON" for a long time. He said, "you said it is better to switch the filament "OFF", if possible, to suppress occurring the microdischarge".

I deeply thought that we must develop the new type of electron gun of high-density alumina, whose outgassing rate is very low, as soon as possible.

We developed a new type of 120-kV electron gun with high-density alumina insulator of extremely low outgassing for JEM-1200EX and later JEMs.

3.1.8 Severe Oil-Vapor Backstreaming, Occurred when the Vacuum System Was Controlled in Error

An operator in the exhibition hall telephoned us just one o'clock that an accident of the oil-vapor backstreaming occurred in a JEM. She explained as follows. "I pressed the button-switch to vent the camera chamber in order to exchange the film magazine. On that moment the chime rang to know the lunch time. So, I pressed the button-switch again to evacuate the camera chamber. Then, the evacuation system was controlled in error, leading to an oil-vapor backstreaming accident. Roughing was skipped, and the high vacuum evacuation valve opened immediately."

I understood what happened. I recognized that the said accident occurred due to the defect of sequence control.

The said accident occurred as follows.

She pressed the button-switch to evacuate the camera chamber again immediately after switching to vent. According to the sequence control, the evacuation valve closed first and then the small vent valve opened about one second after closing the evacuation valve. At that instance the clock chime range to know the lunch time. And she pressed vent switch again to cancel the venting. The Pirani gauge in the camera chamber must judged which valve to be opened, the roughing valve or high-vacuum valve. In the said case, the Pirani gauge judged that the high vacuum valve for the DP-evacuation to be opened due to slow venting and the time lag of a Pirani vacuum gauge. As a result, the evacuation valve for DP opened for the camera chamber with a pressure much higher than the switching pressure of ~10 Pa. Big backstreaming naturally occurred according to the sequence control.

I told her, the operator, as follows. "Sequence control has a defect, in fact. When one presses the button-switch twice with a short interval, the sequence control skips roughing and the evacuation valve for DP opens directly due to a time lag of the Pirani gauge used, causing a big backstreaming accident. Please press the button-switch for the camera chamber more than 3 s later after the first pressing the button-switch."

I informed the detail of the said accident to the electric designing department with the specification of modified sequence control.

Sequence-control of vacuum system having a sufficient time period for thermal-conduction vacuum gauges like a Pirani gauge, received the patent approval [7].

3.1.9 The Lead-Glass Window of the Viewing Chamber, Getting Gloomy After EM-Film Magazine Exchanged

A backstreaming accident occurred on a customer's EM (JEM-100C with the earliest-type cascade DP system). The customer, a researcher in the basic medical sciences, took many sheets of photograph of organism samples, making the montage photo-print of them. In order to make the montage photo-print, the film magazine, containing 48 sheets of film, was exchanged several times a day. JEM-100C was equipped with the viewing chamber of large volume, more than 40 L. Frankly speaking, the fore-pressure of the DP for evacuating the viewing chamber easily rose above its critical fore-pressure, causing severe oil-vapor backstreaming.

Additionally speaking, the wide evacuation pipe from DP faces onto the lead-glass window of the viewing chamber. When oil vapor backstreaming occurred, oil-vapor molecules easily condensed on the surface of the window glass.

I visited the said customer together with a service engineer to resolve the backstreaming problem. I recognized that this backstreaming claim was hard to be resolved. I must explain frankly what happened in the DP system for the camera chamber. And I must propose some ideas to reduce the oil-vapor backstreaming.

I first did an experiment on the backstreaming in the said microscope by inserting a cross-shape plate into the main evacuation pipe for the viewing chamber. And, the film magazine containing 48 sheets of non-degassed film was exchanged as the customer did. We watched the window glass after the viewing chamber was evacuated by the DP routinely. We noticed that the window glass got gloomy a little bit. We exchanged the film magazine containing 48 sheets of non-degassed film again and watched the window glass. After exchanged the film magazine second times, we noticed the cross-shaped shadow picture on the window glass clearly. We, three persons, saw the oil-vapor backstreaming by our eyes.

I explained how to occur the oil-vapor backstreaming, as follows. Just after switching over the evacuation modes, from roughing by a mechanical rotary pump (RP) to a high-speed pumping by a diffusion pump (DP), excess gas loads of space gas and outgassing, flow into the DP-foreline evacuated by the backing RP. When the pressure of the foreline rises above the critical fore-pressure of DP, an abnormal backstreaming occurs.

The means to reduce the backstreaming, which I proposed, were as follows.

1. A buffer tank to be installed to increase the fore-line volume.
2. The switching pressure from roughing by RP to high-vacuum pumping by DP to be set at less than 10 Pa instead of 13 Pa.
3. Degassed film to be used instead of non-degassed film.

Two modified systems are there, I thought, to resolve this severe backstreaming problem, one is to provide an additional DP for evacuating the camera chamber to be evacuated by the DP-DP in-series system, and the other to equip a small-conductance bypass valve (for instance about 5 L/s) which opens several second (for instance 5 s) before the large evacuation valve opens. However, equipping an additional DP to the existing DP system for the camera chamber was practically difficult because of the shortage of the space inside the framework of the microscope. And for the idea of small by-pass valve, the experiment on the by-pass valve was not made at that time. Therefore, I could not mention the above-mentioned two ideas.

We deeply thought we must improve the DP evacuation system for JEMs, as soon as possible.

How to suppress the pressure-rise at the fore-line just after switching-over will be described in Sects. 4.1.6, 4.1.7 and 4.1.8, in details.

3.1.10   "Make JEOL Sputter Ion Pump (Noble-Pump Type) Be Capable to Pump Xe Gas Stably"

The claim was as follows. "The customer using a JAMP (JEOL Auger Microprobe Spectrometer) claims that the JEOL sputter ion pump (SIP) cannot pump the Xe gas stably, accompanying pressure fluctuation. The customer uses the Xe gas for sputtering the specimens for a higher sputtering rate, comparing with Ar-ion sputtering."

As will be described in Sect. 5.2.2, JEOL SIP of noble-pump type used the cathode of "slotted Ta on flat Ti/flat Ti" pair for Ar-gas pumping. Frankly speaking, we expressed my question when hearing Xe gas to be used for sputtering, "why does the customer use Ar gas for sputtering?" The answer is, "the sputtering rate is very high when using Xe gas."

We must respond the customer's claim rapidly.

We had read the important papers, "The physics of sputter-ion pumps" by R. L. Jepsen (see Ref. [1] of Chap. 5) and "Pumping mechanisms for inert gases in diode Penning pump" by P. N. Baker and l. Laurenson (see Ref. [5] of Chap. 5), presented in details in Sects. 5.1.1 and 5.1.5.

"What is the mass-number of Xe?" After checking the mass-number of Xe as 131, which is much higher than the mass number 40 of Ar, we discussed as follows. "The papers by R. L. Jepsen (1968) and P. N. Baker and l. Laurenson (1972) were instructive for us to understand that for Xe gas pumping the noble-type SIP should have high-mass cathode for both sides. "Try the slotted Ta cathodes for both sides."

And we got the information that the customer's SIP with the cathodes of "slotted Ta on flat Ti" for both sides can pump Xe gas stably.

As is well known the Ta plate is much expensive, compared with Ti plate. Therefore, the JEOL standard noble-type SIPs have "slotted Ta on flat Ti/flat Ti" pair, which can be easily exchanged to "slotted Ta on flat Ti/ slotted Ta on flat Ti" pair optionally.

3.2   Information on Working Fluid for DP

3.2.1   Silicone Fluid

Silicone fluid DC-705, working fluid for DP, was widely used in ultrahigh-vacuum DPs in 1970s. DC-705 has the following advantages.

- Conventional DP works well with DC-705 by just increasing the heater power.
- DC-705 is highly resistant to be oxidized when contacting with air molecules in a vacuum field.
- DC-705 has a very low vapor pressure of 10^{-8} Pa range, which was quite attractive for UHV instruments.

A claim naturally occurred as that DC-705 should be applied to DPs for EMs. We had to evaluate such silicone fluid, as well as other hydrocarbon fluids, such as Apiezon fluids, Lion fluid and polyphenylether.

Our experiment on the contamination building up for many kinds of fluid were reported in a Journal in Japanese [8]. Experimental results on various fluid were as follows [8].

The polymerized deposit for silicone grease was the thickest among all fluid tested. The polymerized film was charged up, emitting light when being irradiated with an electron beam. This means that the polymerized film was non-conductive in electricity, which might cause microdischarge on the electron gun.

The polymerized film for DC-705 was also charged up when being irradiated with an electron beam. DC-705 should be avoided to be used for HT electron gun in respect to microdischarge.

Polyphenylether fluid (Santovac-5) and Apiezon-L were barely responsible for the contamination build up in spite of prolonged irradiation time. (The results on polyphenylether were not reported in the paper [8] deliberately.)

Our conclusion was that the DC-705 should not be applied to the DPs for JEMs. And we ascertained the extreme low vapor pressure of polyphenylether at room temperature.

3.2.2   Perfluoropolyether

A paper by Ambrose et al. [9] presented the reduction of polymer growth in an EM by using a fluorocarbon-oxide pump fluid, as follows [9].

The microscope was evacuated by untrapped rotary and diffusion pumps(RP and DP). Initially the RP was charged with a mineral oil (Shell) and the DP (76 mm diameter) with Apiezon B. The average pressure, measured above the DP, was 10^{-5} Torr (10^{-3} Pa). To enhance the effects of polymer growth the anti-contamination shields were left not cooled. A single crystal film of gold was viewed at 100,000× magnification with a focused 100 kV beam. The electron current at the specimen was 3×10^{-7} A and after 15 min exposure to the beam the image on the screen had been obliterated by a carbonaceous deposit growing on

Fig. 3.11 (**a**) Bright-field
micrograph showing the
extent of the contamination
on the specimen when using
hydrocarbon fluid in the
vacuum pumps. (**b**)
Bright-field micrograph
showing the reduction in
contamination of the
specimen when using
fluorocarbon oxide fluid in
the vacuum pumps [9]

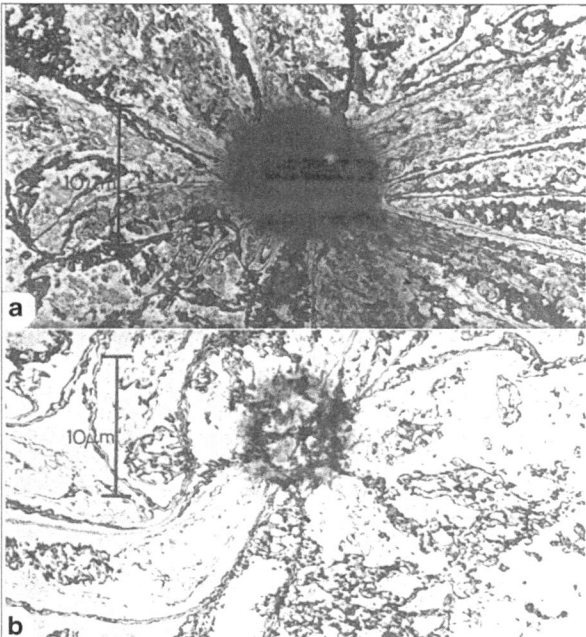

the specimen. By reducing the magnification to 1,000× the extent of the contamination
could be seen in a bright-field micrograph as a black spot surrounded by a cleaner area.

The RP was then replaced by one charged with a suitable grade of fluorocarbon fluid.
The hydrocarbon oil was removed from the DP and after the interior of the pump had been
thoroughly cleaned it was re-charged with a fluorocarbon of sufficient volatility for vapor
stream use. As far as possible the hydrocarbon oil was also removed from associated fit-
tings, such as valves. The anode, screening tube and apertures were cleaned as in a routine
maintenance. The microscope was evacuated and a pressure of $\leq 5 \times 10^{-5}$ was reached on
this and successive pump-downs. After 6 days continuous routine use the contaminant test
described above was repeated using the same gold film and irradiation conditions. After the
15 min bombardment period detail of the specimen image could still be discerned clearly.
When the magnification was reduced to 1,000×, a bright-field micrograph showed that
some darkening of the irradiated area had occurred.

A comparison of the bright-field micrographs at 1,000× made during experimental runs
showed that the area surrounding the irradiated spot was darker when the vacuum pumps had
been filled with organic fluids. This effect could arise from a difference in contamination level
induced by scattered electrons or optical density differences occurring during photographic
processing. Micrographs of the adjacent contamination areas on the gold specimen were
therefore prepared on a single photographic plate using a split screen technique. To avoid
additional contamination and variations in electron beam conditions the micrographs were
taken within 30 s of each other. These micrographs, Fig. 3.11a, b, show that use of fluorocarbon oxide fluids reduces the contamination rate both from primary and scattered electrons [9].

Most of the researchers using EMs naturally read this article because it is presented in *Journal of Microscopy*, which they were familiar with. We felt a kind of
pressure to apply fluorocarbon oxide fluid instead of hydrocarbon fluid for DP and
RP to resolve the contamination problem.

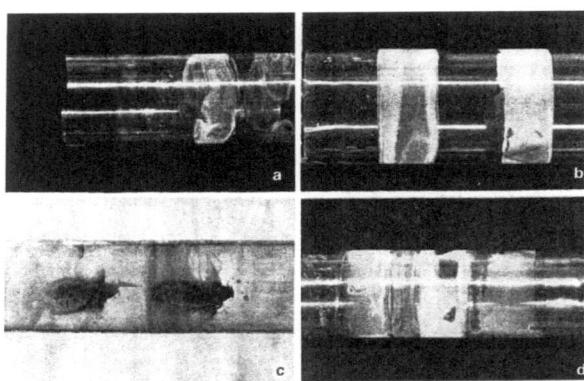

Fig. 3.12 (**a**) Hydrogen discharge, perfluoro- polyether fluid. (**b**) Argon discharge, perfluoropoly-ether fluid (silver film coated on outside of bottle beneath coupling electrodes and subsequently removed before photography). (**c**) Argon discharge, polyphenylether fluid. (**d**) Argon discharge, perfluoropolyether fluid (showing white P.T.F.E.-like deposit which was removed before photographing in (**b**)) [10]

Any vacuum pump with fluorocarbon oxide fluid did not come onto the market at that time. It was questionable whether our DP or RP worked well or not with one of fluorocarbon oxide fluids. It was uncertain whether perfluoropolyether can be cleaned or not with conventional cleaning solvent when an accident of DP-oil-vapor backstreaming occurred in EMs.

Another important paper on perfluoropolyether by Holland et al. [10] was sent us by mail from JEOL USA. Summary are as follows [10].

A study of perfluoropolyether fluids under electron and ion bombardment has shown them to be well suited for vacuum use in the presence of charged particles. A dc-glow discharge was sustained in hydrogen, oxygen, helium, air, argon and Freon 14 between plane electrodes previously smeared with perfluoropolyether. After the discharge both the cath-ode and the anode surfaces remained visually clean except when hydrogen was used as a discharge gas when a polymer-like film was formed on the cathode and a brown film depos-ited on the anode. The brown deposit on the anode was a neutral condensate formed inde-pendently of electron bombardment. For comparison Apiezon C, Silicone 704, Santovac 5 and Dow Corning FS.1265, were also exposed to argon and hydrogen dc-glow discharges, when solid deposits were produced on both anode and cathode. Perfluoropolyether fluid was then smeared in the quartz bottle of an rf ion source and discharges were run in hydro-gen and argon; considerable etching was observed on the inside of the bottle. By compari-son a smear of Santovac-5 in the bottle did not give rise to etching but left a tarry deposit.

The results of the dc glow discharge experiment are shown in Fig. 3.12 [10].

It should be noted that white P.T.F.E.-like deposit was formed when perfluo-ropolyether fluid under Ar discharge.

And the following discussion [10] must be noted.

The experiments described above confirm the resistance of perfluoropolyether fluids to polymerization by electron and ion bombardment, except in the case of a hydrogen discharge when hard tenacious deposits are formed on the cathode electrode. It is believed that in this case a chemical reaction occurs at the cathode surface with the simultaneous evaporation of neutral molecules and very probably the release of hydrofluoric acid vapor. The latter is suggested by the abnormally pronounced etching of the quartz bottle. Direct mass analysis of the reactive components with the rf discharge was not possible because of the problems of extracting into a mass spectrometer ion source. However, mass analysis of the perfluoropolyether vapor was done on a magnetic sector field instrument using an electron impact ion source: both free fluorine and HF were observed in the fragmentation spectrum. It has been concluded that the HF was formed by reaction between fluorine and desorbed water vapor.

A possible reaction is

$[C_3F_6O_{1.1 \text{ to } 1.2}]_n + H_2O \rightarrow$ [polymer] + HF + solid deposit and vapors [10].

We discussed on perfluoropolyether and concluded as follows.

H_2O pressure was quite high in EMs. When the molecules evaporated from perfluoropolyether fluid adhere on the surfaces of the electrodes of the electron gun or the specimen, HF would be generated when irradiated with electron beams, causing a trouble.

Our conclusion at that time was "not so hurry to apply the perfluoropolyether in our microscopes."

3.2.3   Polyphenylether

Polyphenylether (Santovac-5, Monsanto, USA), came onto the market around 1980, which characterized extremely low vapor pressure and high viscosity at room temperature.

The article "the effect of the inlet valve on the ultimate vacua above integrated pumping groups" by Dennis [11] appeared in *Vacuum*, a well known journal. The fluid applied to the integrated pumping group of Edwards (England) was polyphenylether (Santovac-5). We were motivated by the article [11] to develop the diffusion pump which works well with the fluid, polyphenylether.

The schematic of a three-stage integrated pumping group of Edwards is presented in Fig. 3.13. Construction of Edwards DP is quite different from the construction of conventional DPs. It features in the followings.

- Large cold cap, cooled with water
- Large top jet nozzle, having downward angle
- Large cold cap, having downward angle of perpendicular
- Pump vessel, made of thin stainless pipe

We discussed as that we JEOL must develop the DP, working well with polyphenylether. The JEOL-made DP with polyphenylether is described in Chap. 4.

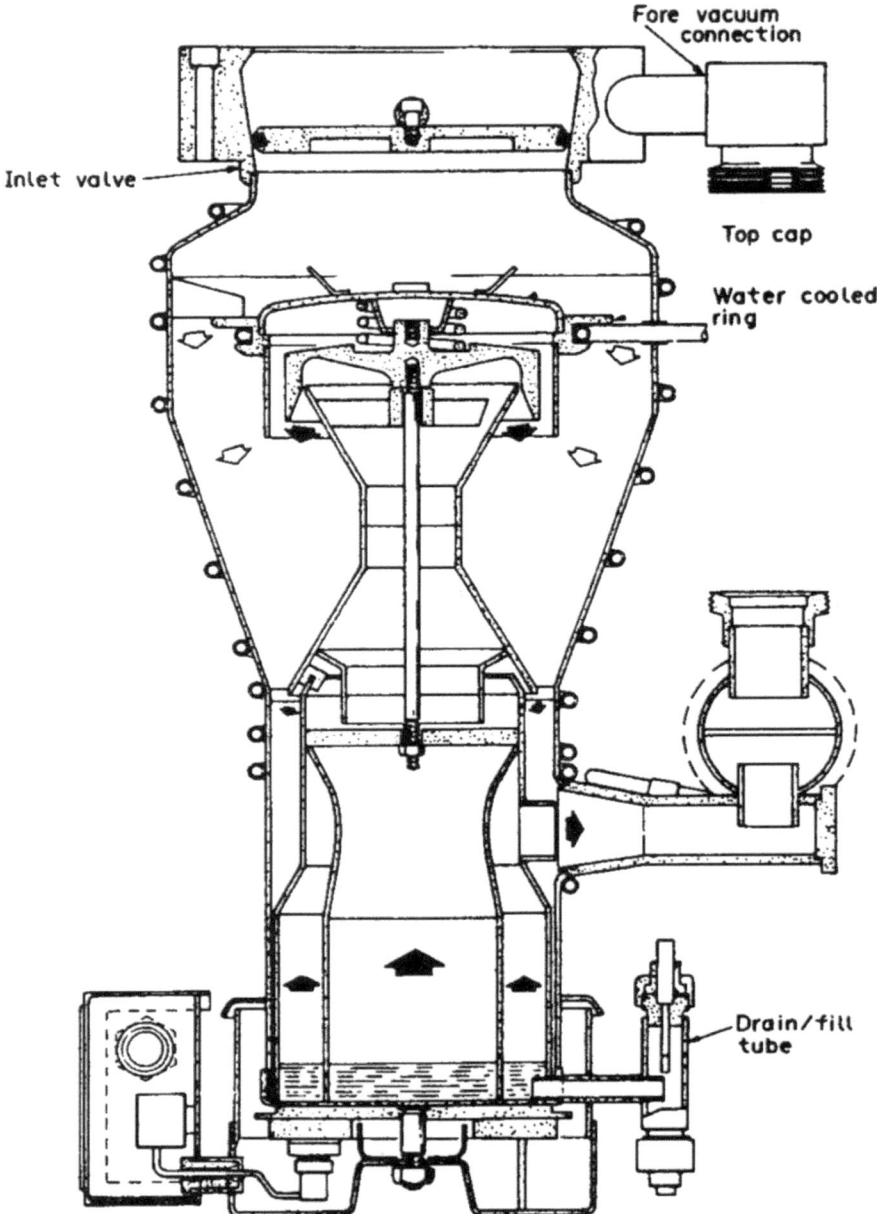

Fig. 3.13 Schematic of a three-stage integrated pumping group [11]

References

1. Yoshimura N (1990) Discussion on methods for measuring the outgassing rate. J Vac Soc Jpn 33(5):475–481 (in Japanese)
2. Takada T, Yoshimura N, JEOL Co. Ltd (1984) Patent no 1984–35,502, 29 Aug 1984
3. Heide HG (1962) The prevention of contamination without beam damage to the specimen. In: Fifth international congress for electron microscopy A-4
4. Yoshimura N (1989) Water vapor permeation through Viton O ring seals. J Vac Sci Technol A 7(1):110–112
5. Hablanian MH, Maliakal JC (1973) Advances in diffusion pump technology. J Vac Sci Technol 10(1):58–64
6. Watanabe H, Yoshimura N, Katoh S, Kobayashi N (1987) Microdischarges on an electron gun under high vacuum. J Vac Sci Technol A 5(1):92–97
7. Yoshimura N, Kobayashi N, JEOL Co. Ltd (1984) Patent no 1984–35,502, 18 May 1984
8. Yoshimura N, Oikawa H (1970) Observation of polymerized films induced by irradiation of electron beams. J Vac Soc Jpn 13(5):171–177 (in Japanese)
9. Ambrose BK, Holland L, Laurenson L (1972) Reduction of polymer growth in electron microscopes by use of a fluorocarbon oxide pump fluid. J Microscopy 96(Pt 3):389–391
10. Holland L, Laurenson L, Hurley RE, Williams K (1973) The behaviour of perfluoropolyether and other vacuum fluids under ion and electron bombardment. Nucl Instrum Methods 111:555–560
11. Dennis NTM, Laurenson L, Devaney A, Colwell BH (1982) The effect of the inlet valve on the ultimate vacua above integrated pumping groups. Vacuum 32(10/11):631–633

Chapter 4
Development of the Evacuation Systems for JEMs

Abstract The evacuation systems and their pumping devices for JEMs were improved year by year, in order to achieve very clean vacuum. We examined the DP cascade system thoroughly in respect of DP oil-vapor backstreaming, and refined the DP-stack and the system year by year according to experimental results. And finally, we established the refined cascade DP system for JEMs.

4.1 DP Evacuation Systems for JEMs

DP in-series system can create a clean vacuum. The differential evacuation system is essential in order to obtain a clean vacuum in the specimen chamber. For creating the extremely low pressure, sputter ion pumps (SIPs) with relatively strong magnet are essential.

4.1.1 Backstreaming of Rotary-Pump Oil Vapor

Holland examined the Effects on RP backstreaming rate of conductance, gas flow and pressure (1971) [1].

Pipe Conductance [1]

The rate of backstreaming from a rotary mechanical pump (RP) will be limited by the conductance of the connecting pipe-lines and baffles if the backstreaming occurs under molecular flow conditions.

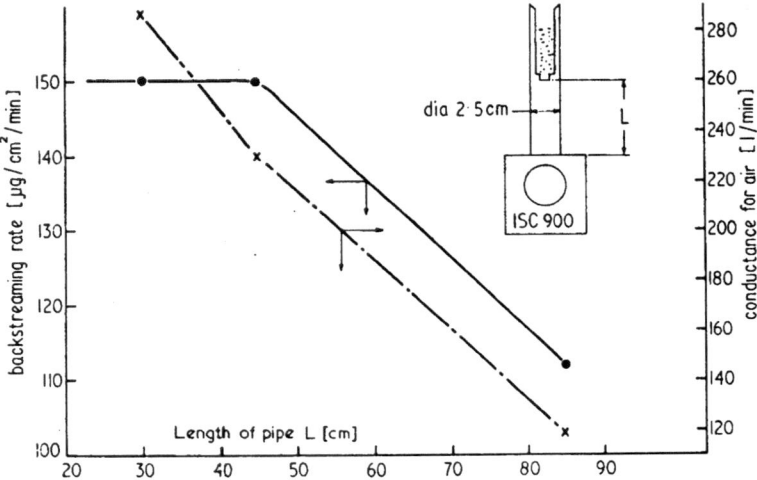

Fig. 4.1 Backstreaming rate as function of pipe conductance 900 L/min single-stage RP [1]

	P₁ (Pirani) (Torr)	P₁ (Pirani) (Torr)	B.S. rate (μg/cm²/min)
Table 4.1 Air inlet through gas ballast valve (static system) [1]	0.010	0.007	36
	0.030	0.025	10.4
	0.050	0.035	7.2
	0.070	0.060	5.2
	0.1	0.085	4.8
	0.2	0.2	2.9
	0.3	0.3	2.2

Shown in Fig. 4.1 is the backstreaming rate as a function of a pipe conductance of a single-stage 900 L/min RP charged with Edwards No 16 oil. The figure shows that at the pump ultimate pressure ($\sim10^{-2}$ Torr) the backstreaming rate was proportional to the pipe conductance which was varied by changing its length.

Gas Pressure and Flow [1]

The backstreaming rate of a pump is reduced at high ultimate pressures because the gas molecules hinder the flow of oil components from the pump. Thus a single stage pump which is gas ballasted and has an ultimate pressure >0.1 Torr can give a low backstreaming rate than a two-stage RP with a low ultimate pressure.

A study is being made to determine the effect of gas pressure when raised by gas-ballasting and flow into the pump. Given in Table 4.1 are the backstreaming rates measured for a 75 L/min RP charged with a mineral oil when air has been admitted via the gas-ballast valve (static gas) as in the schematic drawing (Fig. 4.2)

The arrangement used for admitting air to the pump to provide a gas flow counter to that of the backstreaming is shown in Fig. 4.3. The gas was admitted at 63 and

Fig. 4.2 Experimental set-up for measuring the data of Table 4.1 [1]

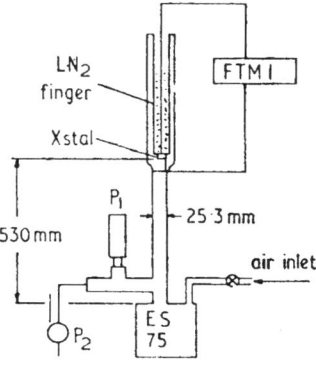

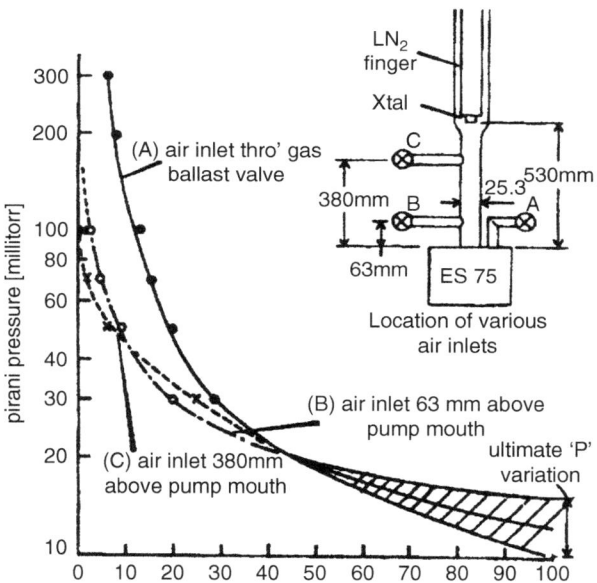

Fig. 4.3 Backstreaming rates as a function of pressure during gas flow into a 75 L/min rotary pump charged with No 16 Oil. The air inlets are in the pump line. Rates measured with a stationary gas in the same pressure range are plotted for comparison [1]

380 mm from the pump and the pressure difference across the pipe-line measured during flow to ensure that the pressure measured under static and flow conditions were comparable. The pressure difference during flow did not exceed 5 % of the mean pressure. However, the backstreaming rates plotted in Fig. 4.3 as a function of pressure during gas-ballast and air flow show that when the pump exhausts gas at its normal flow rate at 0.1 Torr the backstreaming rate is reduced to a negligible amount, whereas at this ultimate pressure with gas-ballasting the backstreaming rate is 10 % of the maximum value. When the gas pressure is below about 2×10^{-2} Torr its

Fig. 4.4 Typical rough
evacuation systems [2]

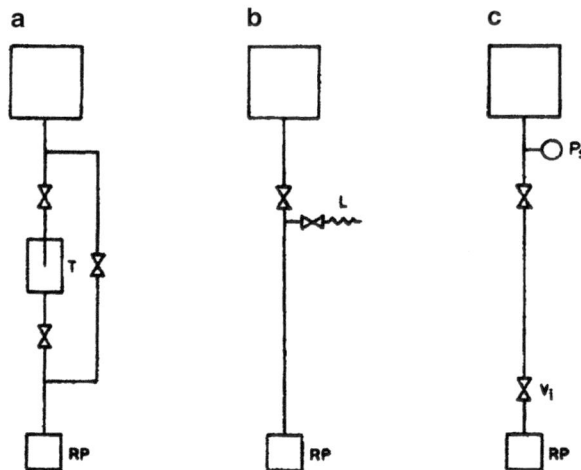

adjustment by gas-ballast or pipeline flow becomes difficult, because the ultimate pressure (Pirani gauge) of a single-stage pump can be easily vary daily from 10^{-2} to 2×10^{-2} Torr even under clean operating conditions. Also, as will be shown below, the pressure of the gas and backstreaming components along the pipe will not only be a function of the individual flow rates and the pipe conductance, but also of their momentum exchange either in the direction of the pump with gas flow or in the direction of a condenser with backstreaming flow.

We discussed and evaluated the RP roughing systems with respect to backstreaming of RP oil vapor and the simplicity of the system, as follows (1984) [2].

Various systems have been proposed to minimize the oil vapor backstreaming from a running RP. Typical systems are presented in Fig. 4.4. The system (a) with a by-pass valve line with a sorption trap T had the disadvantage of practical complexity in operation and is expensive. The system (b) with deliberate leakage L has a disadvantage because the leakage unnecessarily prolongs the roughing time. The system (c) with an additional isolation valve V_i, located just above a running RP, has been evaluated to be the most reasonable system for electron microscopes (EMs). In this system, V_i is opened in the roughing mode only, and closed when the pressure in the chamber reaches a switching pressure P_s which is above 13 Pa. Thus, most of the entire roughing pipe line is not evacuated below P_s with the continuously running RP, resulting in a maintenance-free clean system.

4.1.2 The Influence of Fore-Vacuum Conditions upon Pumping Performance of DP Systems

Hengevoss and Huber [3] investigated the effect of fore-vacuum conditions upon the ultimate pressure using a DP in-series system.

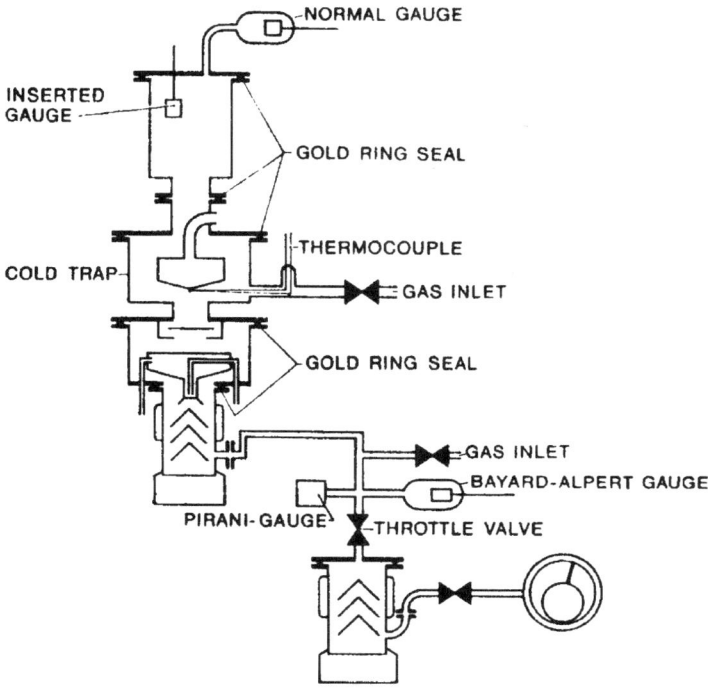

Fig. 4.5 Experimental apparatus. Metal pump system with creep baffle and cold trap. Ultimate vacuum without refrigeration about 1×10^{-9} Torr. Ultimate vacuum with refrigeration about 5×10^{-10} Torr [3]

The Influence of Back Diffusion [3]

The experiments were made with a metal UHV pumping system (Fig. 4.5). Two three-stage fractionating oil diffusion pumps connected in series are used. As driving fluid, silicon oil DC 704 is used. Each pump has a nominal pumping speed of 60 L/s. In the following experiments the cold trap was not cooled.

To investigate the influence of the back diffusion on the ultimate pressure, various gases were passed separately into the intermediate vacuum while the partial pressures in the high vacuum were measured with the mass spectrometer. Figure 4.6 shows the results. On the abscissa the total pressure of the different gases in the intermediate vacuum is plotted, on the ordinate the corresponding partial pressures in the high vacuum. The pressures are calibrated in nitrogen equivalents (N_2-equ.). In the intermediate vacuum the pressures below 10^{-2} Torr are measured with a Bayard-Alpert gauge, pressures above 10^{-3} Torr are measured with a Pirani gauge. In the overlapping region the Pirani gauge was calibrated against the ion gauge, so that the whole pressure scale of the abscissa corresponds to nitrogen equivalents.

Figure 4.6 shows that all gases diffuse from the intermediate vacuum into the high vacuum. As could be expected, the light gases H_2 ($M=2$) and He ($M=4$) show

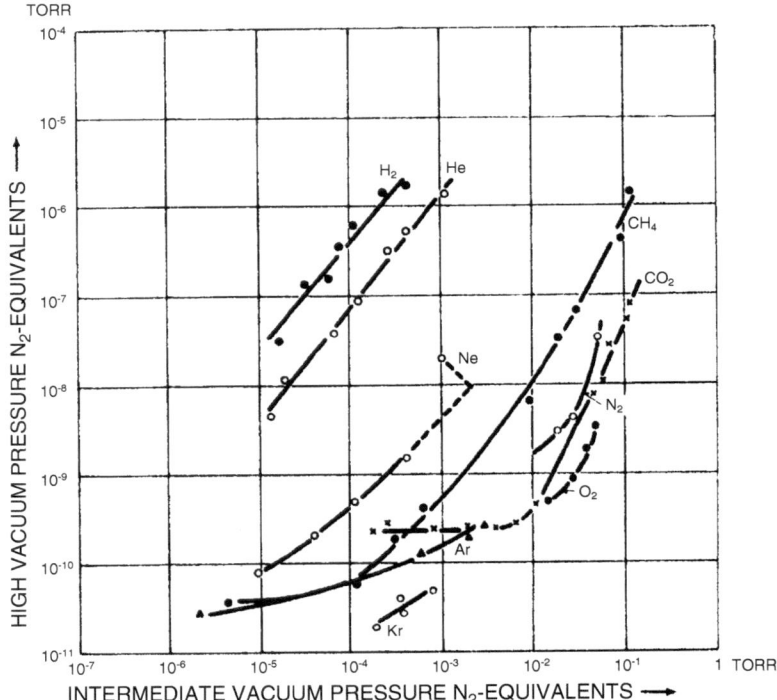

Fig. 4.6 Back diffusion from the intermediate vacuum into the high vacuum. The various gases are passed separately into the intermediate vacuum and the partial pressure of each measured in the high vacuum [3]

the highest diffusion rate. For the inert gases Ne ($M=20$), Ar ($M=40$) and Kr ($M=84$) the diffusion rate apparently decreases in inverse proportion to the molecular weight. The curves for the other gases do not seem to show this dependence with the same accuracy, e.g. it might be expected that the CH_4 curve ($M=16$) would lie higher than the Ne curve ($M=20$) but as seen in Fig. 4.6 the opposite is the case. The reason for this is not quite clear. Possibly solution effects of the gas in the diffusion pump oil are responsible for this deviation from the predicted behavior.

The following rules for production of UHV can be derived from the examination of Fig. 4.6.

In order to produce an ultimate pressure of 1×10^{-9} Torr, the partial pressure of the light gases whose molecular weight lies between 1 and 20, must not be higher at the fore-vacuum side of the main diffusion pump than about 2×10^{-4} Torr N_2 equivalent. In order to produce an ultimate pressure of 10^{-10} Torr, the partial pressure of these gases must not be higher than 1×10^{-5} Torr N_2 equ. At this point it must be noted that water vapor is one of the lighter gases which limit the ultimate pressure by back diffusion. This is an important fact because water vapor is one of the most abundant components of the fore-vacuum atmosphere, if the fore-vacuum part of the pumping system is not baked, which is the normal case.

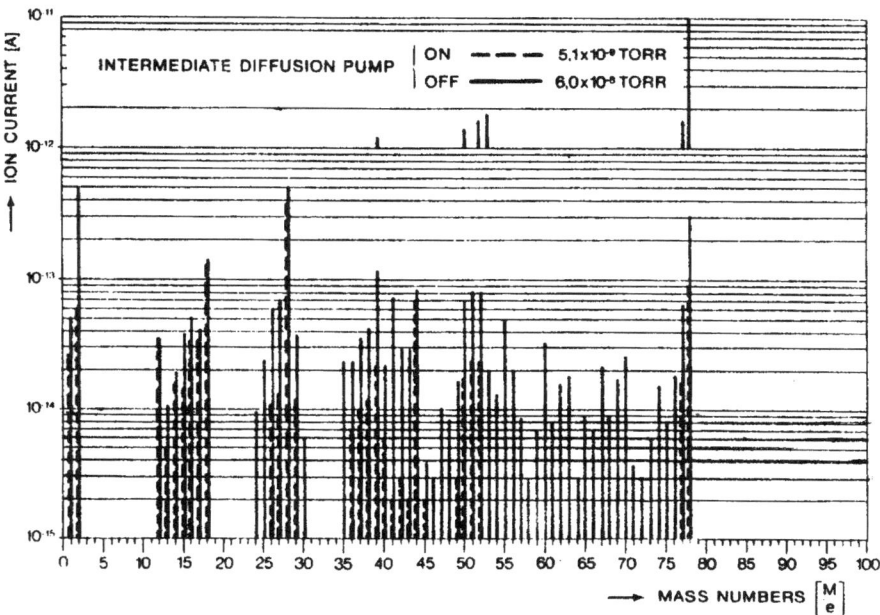

Fig. 4.7 Turning off the booster diffusion pump. The six typical peaks of benzene are noted [3]

Turning Off the Booster Diffusion Pump [3]

To estimate the influence of the booster diffusion pump on the ultimate pressure, this pump was turned off. As a result, the total pressure increased from 5×10^{-9} Torr to 6×10^{-8} Torr. Figure 4.7 shows the spectrum of the remaining gases before and after the pump operation had been stopped. Only the mass lines up to 78 are shown. It was observed that after the booster diffusion pump was turned off, particularly the mass peaks 2 (H_2), 15 (CH_4) and 78 increased, and also 28 (CO) and 44 (CO_2) both increased a little. It is especially noteworthy that after the booster diffusion pump had been turned off, many new peaks appeared in the mass spectrum.

Experimental results show that the two-DP in-series system creates a very clean, ultrahigh vacuum.

The paper by Hengevoss and Huber [3] taught us the following important knowledge: DP in-series system can create the very clean vacuum when using working fluid of very low vapor pressure at room temperature.

4.1.3 DP System with a LN₂ Trap at the Junction of the Main Pumping Line and the Line from the Viewing Chamber

This evacuation system was realized in JEM-100B, which was presented 1973 in details [4].

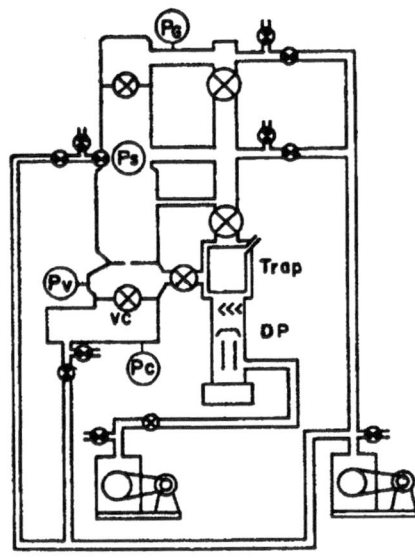

Fig. 4.8 Schematic diagram
of the evacuation system of
JEM-100B [4]

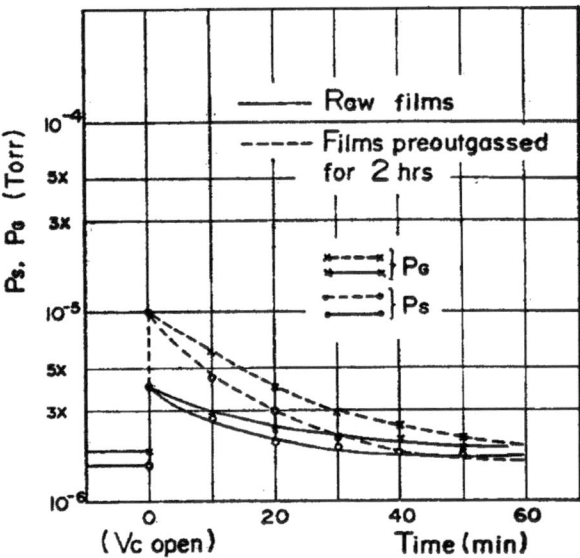

Fig. 4.9 Variations of P_S and
P_G in the system of Fig. 4.8
when the valve V_C is opened
at 0.1 Torr. The trap is being
filled with liq. N_2 [4]

    The evacuation system is shown in Fig. 4.8. The variations of the pressure P_S in
the specimen chamber and P_G in the gun chamber when the airlock valve V_C is
opened at about 0.1 Torr, is presented in Fig. 4.9.

    Thanks to this vacuum system, the pressures in the specimen chamber and the
gun chamber reached the lower 10^{-6} Torr region, when using liq. N_2.

    The evacuation system, equipped with a liq. N_2 trap at the junction of the main
pumping line and the line from the viewing chamber, was registered as a patent [5].

Fig. 4.10 A new evacuation system (the first cascade DP system) for an EM [6]

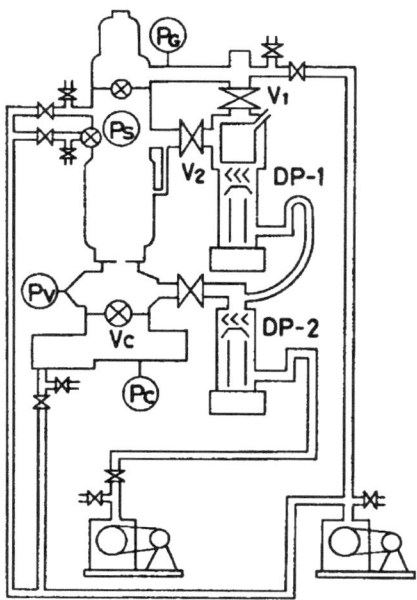

4.1.4 First Presentation of a Cascade DP System

I orally presented the cascade DP system as "a new vacuum system for an electron microscope" at 6th Internal Vacuum Congress 1974 [6]. After then, the typical JEMs (JEM-100C, -100CX, -200CX, 1200EX, -2000EX, and so on) were provided with a type of cascade DP system. The cascade DP system and its DP stack were improved year by year after then.

The system consists of two oil diffusion pumps connected in cascade. The gun chamber and the specimen chamber were evacuated by the upper pump, and the viewing chamber and the camera chamber by the lower one. The system is shown in Fig. 4.10.

Figure 4.11 shows variations of P_S and P_G when V_C is opened after loading the film magazine containing 24 sheets of non-degassed film. Evacuation characteristics after exchanging films or specimens were effectively improved by this system. As a result, the pressures in the specimen chamber and the gun chamber rapidly reached the lower 10^{-6} Torr region.

Features of this new vacuum system are listed below:

• Gases in the viewing chamber and the camera chamber do not adversely affect the vacuum of the gun chamber and the specimen chamber.
• Effective pumping speed of the upper pump is high.
• Fluid in the upper pump is not contaminated by the rotary pump oil.
• This vacuum system is suited for EMs in structure and size.

With this new cascade DP system, appreciable deterioration in vacuum does not occur after exchanging the film magazine.

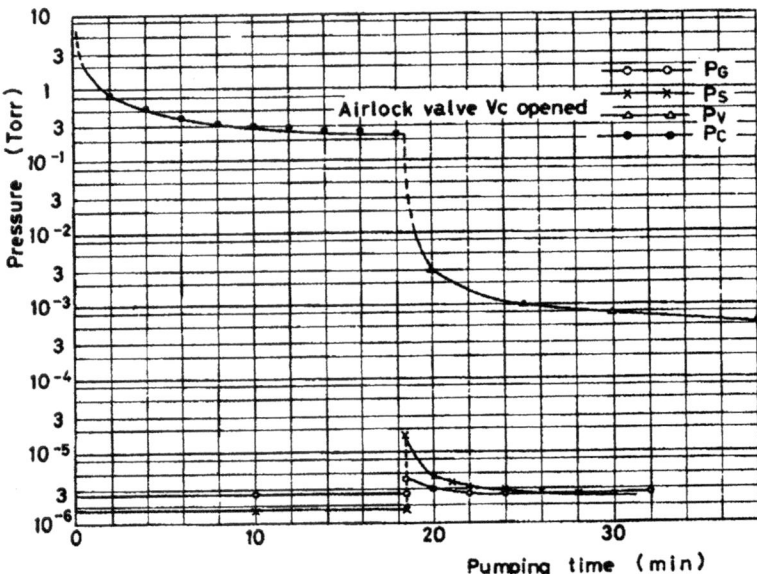

Fig. 4.11 Variations of P_S and P_G when V_C is opened after loading 24 sheets of raw (non-degassed) film [6]. *Note*: liq. N_2 is not used for the trap

4.1.5 Developing the New JEOL DP Stack

The DP stack is composed of the liq. N_2 trap, the chevron-type water-cooled baffle, and the DP itself. The JEOL DP stack was improved year by year around 1980 in order to improve the evacuation systems of JEOL scientific instruments such as EMs and SEMs.

The DPs for JEMs in 1960s were not equipped with an ejector jet nozzle, resulting in their relatively low critical fore-pressure. DP working fluid were Octoil-S [7] or Apiezon-C [7] until the fluid Lion-S [7, 8] (hydrocarbon) was applied to JEOL DP.

We had to design a new diffusion pump for Santovac-5 (polyphenylether [8]). The construction of the pump, intended for Santovac-5, was similar to the pump at the middle of Fig. 3.8. The pump was equipped with baffle plates inside the fore-vacuum-side ejector pipe which was water-cooled, in order to prevent flowing out of fluid. The pump vessel of the first designed pump for Santovac-5 was made of a thick-wall (~3 mm thickness) stainless-steel pipe of 4-in. diameter.

The water-cooled cold cap, equipped for the top jet to reduce the oil-vapor back-streaming, was essential to clean vacuum. However, it was quite difficult for us to design the pump with a cold cap cooled with water. We resolved this problem by designing the new chevron-type baffle with a cold cap. A cold cap machined of copper block was soldered to the center of the chevron-type baffle fins to fit the top jet when the baffle was stacked on the new diffusion pump. The cold cap applied is similar to the type (c) shown in Fig. 4.12.

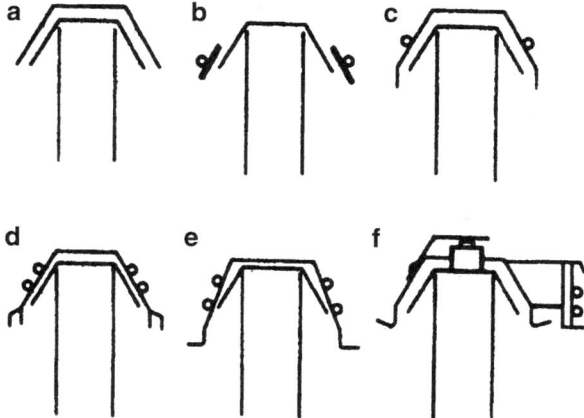

Fig. 4.12 Illustration of the progress in cold cap design [9]. (**a**) J. Ruf and O. Winkler, in Ergebnisse Der Hochvakuumtechnik, edited by M. Auwarter (Wissenschaft. Verlag, Stuttgart, 1957), p. 207. (**b**) B. D. Power and D. J. Crawley, Vacuum 4, 415 (1954) (published 1957). (**c**) M. H. Hablanian and H. A. Steinherz, Trans. Amer. Vac Soc. (Pergamon, New York, 1961), p. 333. (**d**) N. Milleron and L. Levenson, Trans. Amer. Vac. Soc. (Pergamon, New York, 1960), p. 213. (**e**) S. A. Vekshinsky, M. I. Menshikov, and I. S. Rabinovich, Pro. First Internat'l. Cong. Vac. Techniques (Pergamon, London, 1958). (**f**) M. H. Hablanian, J. Vac. Sci. Technol. 6, 265 (1969)

The DP-baffle stack using Santovac-5 (polyphenylether) was examined by varying the heater power around 600 W. However, experimental results were unsatisfied because the recorded vacuum pressure accompanied with many pressure pulses.

We had to modify the pump design. But, we could not answer instantly why such pressure pulses were accompanied with.

Illustration of Fig. 3.13 shows the two different points from our new DP, (1) top jet nozzle, having downward angle, and (2) pump vessel, made of thin stainless-steel plate. The pump of Fig. 3.13 operates well actually.

We discussed as follows.

The vapor pressure of Santovac-5 at room temperature is extremely low, so the temperature of the oil vapor jet impinging the inner surface of the pump-vessel wall must be very high. In order to operate stably, the inner surface of the pump vessel wall must be cooled effectively by water cooling. Apparent difference between the Edwards DP and our new pump (JEOL) is the thickness of the pump-vessel wall. The thickness of Edwards DP may be ~1 mm thickness, and on the other hand, that of the JEOL DP is ~3 mm thickness. This difference must affect the cooling efficiency for the inner surface of the pump-vessel wall. However, as discussed in Sect. 3.1.6, the thick wall of the pump vessel is desirable for anti-vibration effect.

We manufactured the second new pump of steel pipe, nickel plated, instead of stainless-steel pipe. The steel DP/baffle with a cold cap stack using Santovac-5 was examined by varying the heater power around 600 W. Experimental results were satisfactory because any pressure pulse was not accompanied with the recorded vacuum-pressure line.

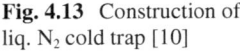

Fig. 4.13 Construction of
liq. N$_2$ cold trap [10]

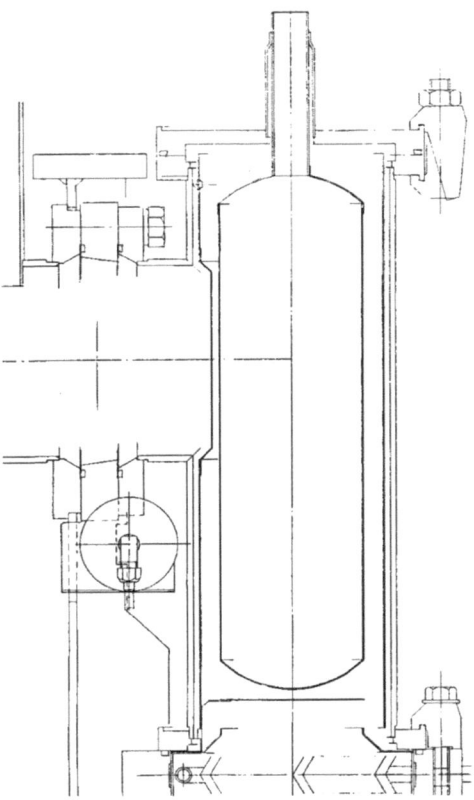

We considered as follows. Thermal conductivity of stainless steel (SS304) is
16.7 W/(mK) and that of steel as high as 99 W/(mK). Therefore, the steel wall with
~4 mm thickness is equivalent to the stainless steel wall with ~1 mm thickness in
respect of thermal conductivity for water cooling.

This newly designed DP of steel pump vessel, Ni plated, worked well to create a
very clean UHV, as shown in Fig. 3.7.

The device installed on the water-cooled baffle was the liq. N$_2$ trap. As discussed
in Sect. 3.1.5, the temperature of the cold surface of the trap is desired to be raised
to room temperature at least once a week with the JEM being switched "OFF". In
order to realize this condition the liq. N$_2$-holding time of the trap is desired from 15
to 20 h. When JEMs are equipped with such a trap with a long holding time for liq.
N$_2$, users of JEMs do no necessitate an automatic liq. N$_2$ supplying device. Liq. N$_2$
is still held next morning in the JEM being switched "ON" when the trap is filled
with liq. N$_2$ at the evening of weekday. And, the trap will be empty during the week-
end holidays in the electron microscope being switched "OFF".

The cross section of the newly designed liq. N$_2$ trap is shown in Fig. 4.13 [10]. This
trap is equipped with a thermal reflecting thin-wall cylinder in order to prolong the liq.
N$_2$ holding time. The liq. N$_2$ bottle and the thermal reflecting cylinder can be removed
from the main pumping pipe for the customers who do not desire using the liq. N$_2$.

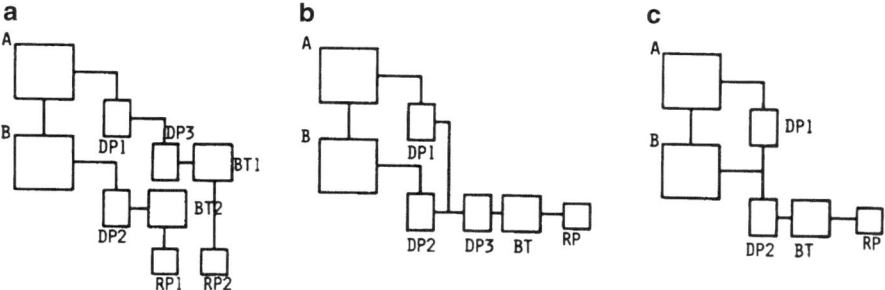

Fig. 4.14 Differential DP systems for TEM. (**a**) Independent system, (**b**) parallel system with a common backing line, and (**c**) cascade DP system [2]

Next, we must develop the further refined cascade-DP evacuation system, into which the newly designed DP stack using Santovac-5 is to be installed.

4.1.6   Refined Cascade DP System

The cascade DP systems of JEMs in the middle of 1970s have several defects. As described in Sect. 3.1.9, backstreaming of DP oil vapor just after exchanging the film magazine might be the most serious defect.

We examined the DP cascade system thoroughly in respect of DP oil-vapor backstreaming, and refined the system and the DP-stack year by year according to experimental results. And finally, we established the refined cascade DP system, which was reported 1984 [2]. The newly modified DP-stack was applied to the refined system, which has been described in Sect. 4.1.5.

The residual gas spectrum in an EM with the refined cascade DP system indicates a very clean vacuum with hydrocarbon partial pressures of less than 3×10^{-7} Pa [2].

The whole column can be considered to be divided into two parts by a small-conductance aperture positioned just below the projector lens pole-pieces. Typical differential evacuation systems, (a) independent system, (b) parallel system with a common backing line, and (c) cascade system, are presented in Fig. 4.14, where the chamber A corresponds to the column, and B to the camera chamber. All systems have a DP-DP in-series system for the chamber A to obtain a clean vacuum and a buffer tank BT for proper transitional evacuation. These systems are evaluated with respect to correlation (mutual relationship) in gas flow between the chambers A and B.

Consider a large leakage occurring in the chamber A in the system (b). Leaked gases are compressed immediately into the common foreline, and flow backwards with dense oil vapor into the chamber B through DP2 when P_b rises above the critical pressure, P_c. Leakage in the chamber B also causes similar backstreaming into the chamber A through DP1. On the other hand, this kind of backstreaming does not occur in the independent system (a). Correlation in gas flow is therefore related to system resistance to emergency due to leakage. In the cascade DP system (c), leakage in the chamber A does not cause such backstreaming of oil vapor into the chamber B, whereas leakage at B may cause backstreaming into A through DP1. In the operation of a TEM, leakage may occur in the column, and almost never at the camera chamber, because most movable devices, such as the specimen holder and apertures, are located in the column, associated with the chamber A. Considering such actual circumstances, correlation in gas flow between A and B is evaluated as tolerable for the system (c). For system cost and occupied space, the cascade DP system (c) is best, which can be easily recognized by comparing the number of system parts with each other.

The cascade DP system with BT is clearly evaluated to be the most practical and reasonable for a TEM, where the column is evacuated by DP1 backed by DP2, and the camera chamber by DP2 with a large-volume foreline backed by RP.

Safety Systems of the Refined Cascade DP System [2]

DP systems must be basically controlled by the backing pressure P_b detected by a gauge at the foreline of DP. When P_b rises to the reference pressure P_r which is set to be lower than the critical backing pressure P_c in any evacuation mode, the vacuum system is rapidly driven to a safe state. This system is called the "basic" safety system here. It serves well against the following emergencies:

- Leakage at any part of the whole column
- Leakage at any part of the DP system
- Failures of the backing RP
- Failures of gauges for switching the evacuation mode

In addition, the vacuum system is protected with the "normal" safety system which drives the system to a safe state using signals which have detected failures on system parts such as the power supply including fuses, cooling water system, pneumatic air system for valves, and DP heaters. The "normal" safety system is activated in all evacuation modes including warming up.

Moreover, the cascade DP system is provided with a safety valve which is closed temporarily, regardless of the existing failures, to protect the vacuum system just before switching the evacuation mode. This system is called the "standby" safety system. The system is presented in Fig. 4.15a, where V_s is the safety valve. Consider that air is admitted into the chamber B while the chamber A is continuously evacuated by DP1. First, consider a system without the safety valve V_s on the foreline of DP1. If sealing of V2 is insufficient to isolate the chamber B, the introduced air rapidly flows backwards through DP1 into the chamber A, resulting in serious backstreaming. Similar trouble may occur if P_h in the chamber B is incorrectly detected in switching the evacuation mode. In the cascade DP system with the "standby" safety system, V_S is closed temporarily just before venting the chamber B and switching the evacuation mode. With the association of the "basic" safety system, the evacuation system is rapidly put into a stop state when P_b rises above P_r, so the "standby" safety system protects the vacuum system against the above-mentioned faults. The flow sequences of this system are given in Fig. 4.15b, one is for venting of the chamber B and the other for evacuation of the chamber B.

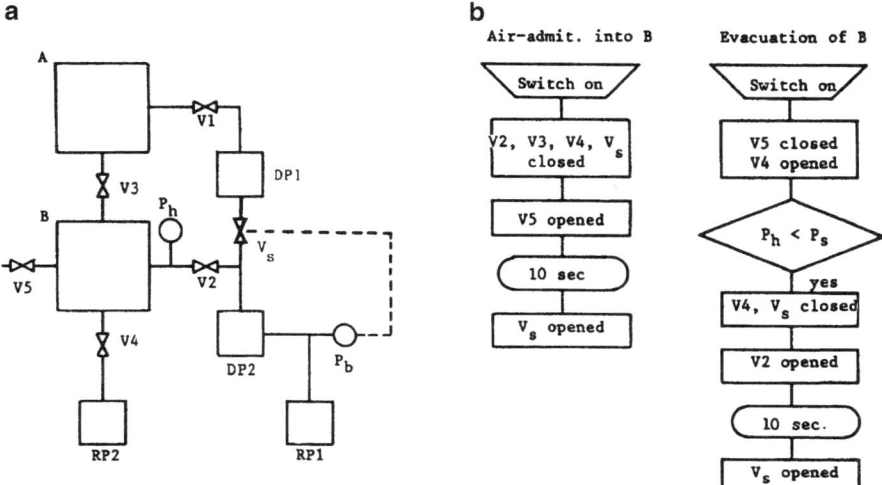

Fig. 4.15 "Standby" safety system. (**a**) System diagram and (**b**) sequence flows: one for venting of the chamber *B* and the other for evacuation of the chamber *B* [2]

Figure 4.16 presents the vacuum system diagram of JEM-1200EX, equipped with a refined cascade DP system. The column and the gun chamber are evacuated by DP1 backed by DP2, while the camera chamber is evacuated by DP2 with the tank RT2 backed by RP. The bypass valve V1 with a low conductance is used for the camera chamber. The volume of the tank RT2 is rather small (approx. 10 L) due to the help of the low-conductance bypass valve V1. The tank serves as a buffer in transitional evacuation as well as a reservoir in roughing. Signals from the Pirani gauge PiG4 at the foreline control the "basic" safety system. The valve V2 is a safety valve in the "standby" safety system which is closed temporarily just before the vacuum system is put into the air-admittance state and the switching the evacuation mode to fine pumping for the camera chamber. The valve V3 on the roughing line is opened in the roughing mode only. A cold trap can be installed above DP1 in the main pumping line.

The function of the low-conductance bypass valve V1 is described in details in Sect. 4.1.8.

The "Standby" safety system, presented in Fig. 4.15, was registered as a patent [11].

Was the Cascade DP System for EM Registered as "Patent"?

We naturally offered the cascade DP system for EM as a patent. But the idea was rejected by the reason that the idea that the second DP of a cascade system was used to evacuate the double O-ring space was already registered as a patent.

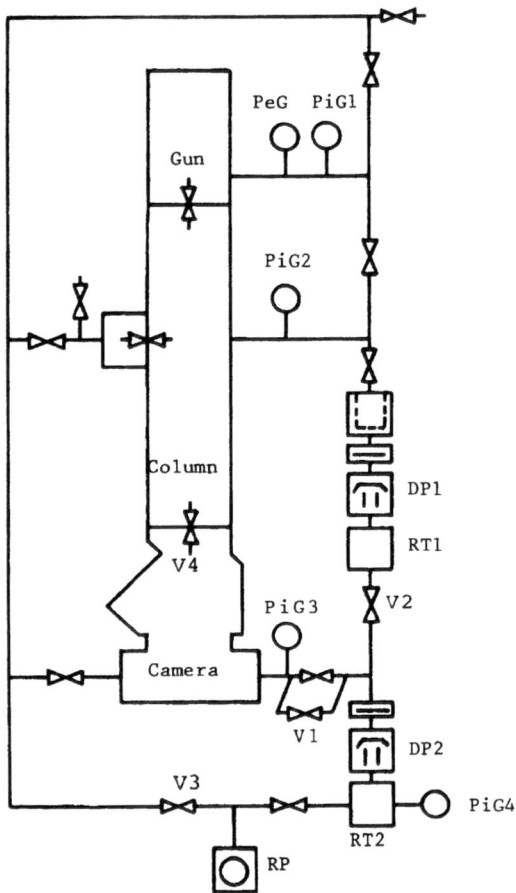

Fig. 4.16 Vacuum system diagram of JEM with a cascade DP system. The liquid-N$_2$ cold trap is optional [2]

4.1.7   DP In-Series System of the JEOL SEM

We reported the performance of the DP in-series system for a wafer-SEM equipped with a large specimen chamber (1991) [12]. The wafer-SEM was developed for inspecting the large-size wafer (6-in. of wafer size), so the large specimen chamber of ~40 L is vented for exchanging a wafer. The title of the paper is "Practical advantages of a cascade diffusion pump system of a scanning electron microscope." In this paper the term "cascade DP system" means "DP in-series system".

DP system can encounter accidental events, such as the breakdown of the DP heater and the cooling water supply trouble.

Considering the actual conditions of accidents, some experiments were conducted on the DP in-series system shown in Fig. 4.17. The DP1 (4 in.) and DP2 (2.5 in.) were both fractionating types with four nozzles including an ejector one,

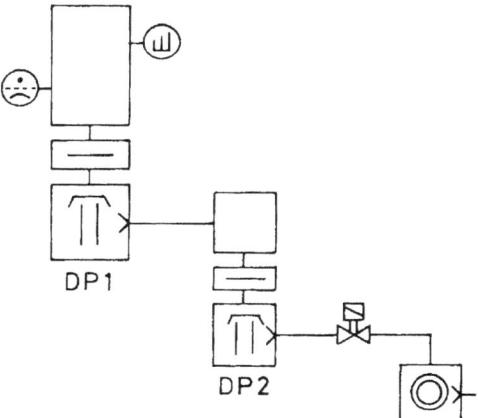

Fig. 4.17   Experimental setup of a DP in-series system [12]

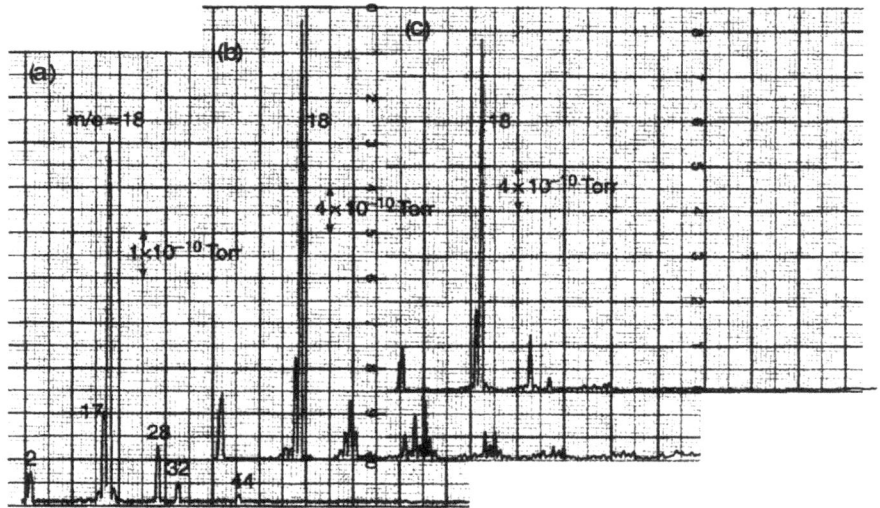

Fig. 4.18   Residual gases depending on the evacuation time t_e after DPs were again switched on in the DP in-series system. (**a**) Before switching off both DPs, (**b**) 4 h of t_e, and (**c**) 26 h [12]

charged with Santovac-5 (polyphenylether). The tolerable forepressure of both pumps was 38 Pa. The Bayard-Alpert gauge (BAG) was a glass-tube type with W filament, operated at 1 mA of emission current. The quadrupole mass-filter (MS, W filament) was operated at 0.5 mA. In the experiment of a single DP system the DP2 was kept off.

The chamber (type 304 stainless steel) was evacuated for a long period, following an in situ bakeout (100°C, 3 days). A residual gas spectrum obtained for the chamber evacuated continuously for 1 month is presented in Fig. 4.18a, showing a

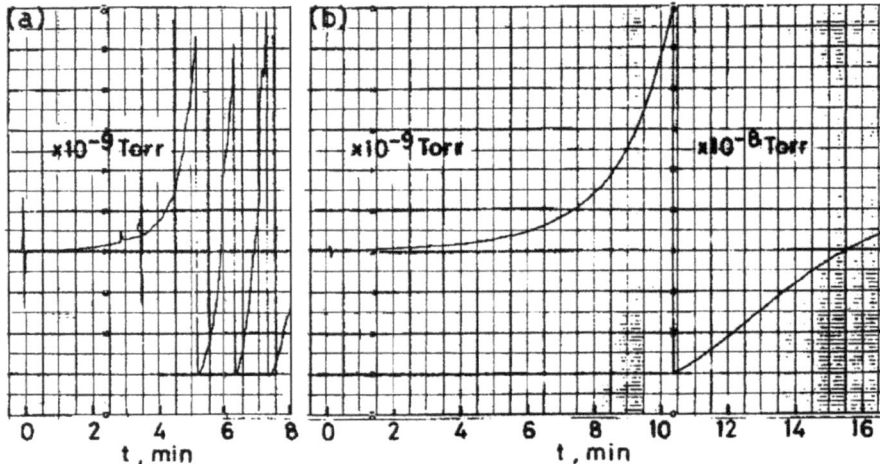

Fig. 4.19 Pressure-rise characteristics after the DP1 was switched off. (**a**) In the DP1-RP system and (**b**) in the DP1-DP2-RP system [12]

very clean vacuum. The spectrum is characterized by the high peaks of water vapor (peaks 17 and 18), which is due to the water vapor permeation through Viton seals. Then, the power to the heaters of both DPs was shut down, and a few minutes later the valve at the foreline of DP2 was closed. After a 1-h cool down, both DPs were simultaneously switched on, just after the valve was opened. The MS (0.5 mA) and BAG (1 mA) were switched on 1 h after supplying DP heater power. Residual gas spectra analyzed 4 and 26 h after switching on the DP heaters are presented in b and c of Fig. 4.18, respectively. A number of hydrocarbon peaks were detected at a short elapsed time after supplying heater power. It takes more than 2 days to recover the clean vacuum without bakeout treatment. Continuous pumping for many days is essential for DP systems to produce a clean vacuum.

Next, only the power to the DP1 heater was switched off on the DP in-series system (DP1-DP2-RP system). The pressure in the chamber rose gradually but smoothly from 4.0×10^{-9} Torr (5.3×10^{-7} Pa) up to an equilibrium pressure of 5.5×10^{-8} Torr (7.3×10^{-6} Pa) produced by the DP2. On the other hand, the pressure on the single DP system (DP1-RP system) rose rapidly to a low vacuum range soon after the power to DP1 was switched off. The pressure rise characteristics of the respective systems are presented in Fig. 4.19a, b. In such a way, the worst situation is avoided with the DP in-series system when the DP1 heater breaks down.

The input voltage to the DP heater often varies.

Residual gases were analyzed by changing the input voltage on the single DP system (DP2 kept off) and on the DP in-series system (DP1-DP2-RP system). The regular input voltage for both DPs is 200 V. The input voltages to both DPs were simultaneously changed from 200 to 180 V and 220 V, and kept constant for 30 min, respectively. Residual gases were analyzed just before changing the input voltage on the respective system, as shown in Fig. 4.20a, b [(1)200 V, (2)180 V, (3)200 V].

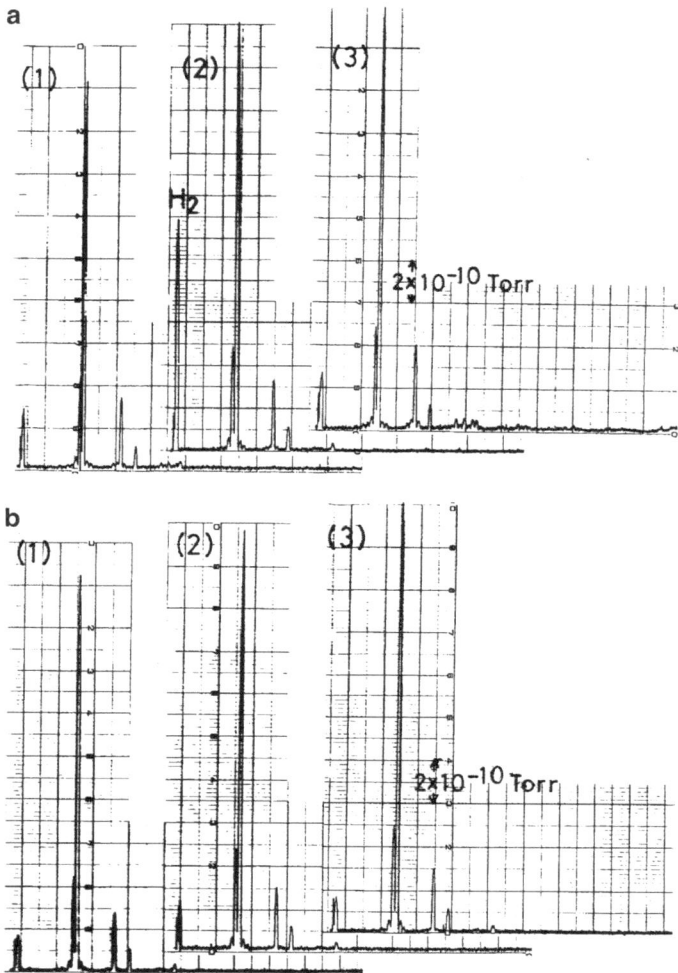

Fig. 4.20 Residual gases depending on the voltage inputs to the DP heaters. (**a**) On the DP1-RP system with (*1*) 200 V, (*2*) 180 V, (*3*) 220 V and (**b**) on the DP1-DP2-RP system with (*1*) 200 V, (*2*) 180 V, (*3*) 220 V [12]

In the single-DP system (a), the peak of H_2 (m/e = 2) increased at 180 V, compared with the peak height of H_2 at 200 V. And at 220 V, hydrocarbon peaks of m/e 39, 41, and 43 increased due to the thermal cracking of working fluid (Santovac-5, polyphenylether).

In the DP in-series system on the other hand, the peak height of every peak did not vary after the input voltage was changed. At 180 V, the limiting compression ratio of the DP1 decreases due to the power reduction. Nevertheless, thanks to the DP2, the H_2 pressure at the foreline of the DP1 was low enough not to increase the

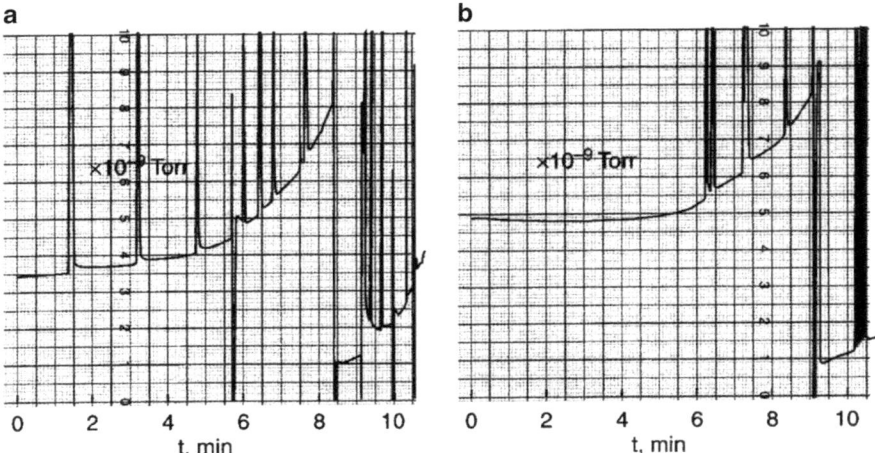

Fig. 4.21 Pressure changes after cooling water was stopped. (**a**) In the DP1-RP system and (**b**) in the DP1-DP2-RP system [12]

H_2 pressure on the high vacuum side. At 220 V, thermal cracking of the working fluid of the DPs must be accelerated due to somewhat higher boiler surface temperature. However, the hydrocarbon cracking products were quickly pumped out by the DP2, with the result that there was no increase of hydrocarbon peaks.

The DP system sometimes encounters accidents in cooling water lines. On both systems, the high-vacuum pressure indicated by the BAG was recorded after the cooling water line to both DPs was shut off. Experimental results are presented in Fig. 4.21, where the pressure range of the recorder chart for both systems was switched to the $\times 10^{-8}$ Torr range at about 9 min of elapsed time.

In the single DP system (a), the gradual pressure rise with large, intermittent pressure pulses occurred 1.5 min after the water line was closed. On the other hand, in the DP in-series system (b), the pressure was kept almost constant for 4 min after the water line was closed, and then the gradual pressure rise with large, intermittent pressure pulses occurred.

Scanning electron microscopes (SEMs) used in the semiconductor industry generally equip large specimen chambers. A SEM with x-ray spectrometers has a large specimen chamber. A clean high vacuum is required in the specimen chamber to minimize specimen contamination. Rapid pumping to high vacuum is also desirable for general efficiency.

In the single DP system with a large chamber the forepressure P_f of DP may rise above its tolerable limit P_t when improperly switching from roughing to fine pumping, resulting in increasing the backstreaming of DP oil vapor. Therefore, the DP system of an instrument of clean high vacuum must be basically controlled by a P_f signal detected at the foreline of the DP. Nevertheless, the system tolerance to overload is necessary.

Fig. 4.22 Vacuum system diagram of SEM equipped with a DP in-series system [12]

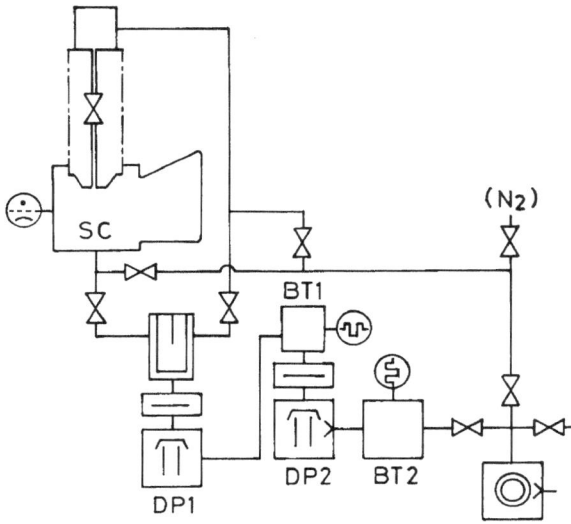

A DP in-series system with a buffer tank (BT) meets the requirements for the SEM vacuum system mentioned above. The vacuum system diagram of a SEM with a DP in-series system is presented in Fig. 4.22. The specimen chamber equips three X-ray spectrometers, having an effective volume of 42 L. The chamber also equips the standard movable mechanisms, such as a large specimen stage and a specimen exchanging mechanism. Such movable mechanisms were lubricated with Apiezon grease, showing considerable outgassing when evacuated after being exposed to the atmosphere. The DP1 (4 in.) and DP2 (2.5 in.) were charged with Santovac-5. The volumes of BT1 and BT2 are 1 and 10 L, respectively. The trap above the DP1 was not cooled throughout the following experiments. For the experiments on the SEM with the single DP system, the DP2 and BT1 were removed, and the foreline of the DP1 was directly connected to the BT2. Note that the pumping speed of the mechanical rotary pump (RP) was rather small (100 L/min) for the large specimen chamber.

On the respective systems the fore-pressures P_f of the DP1, indicated by a Pirani gauge, were recorded when the evacuation mode was switched. The P_f on the single DP system (a) was measured at the BT2, and the P_f on the DP in-series system (b) was measured at the BT1. Experimental results showing the maximum P_f, as a function of switching pressure P_s, are presented in Fig. 4.23.

On the single DP system the maximum rise of P_f was much higher than that on the DP in-series system, although the volume of the BT2 (10 L) was much larger than that of the BT1 (1 L).

The tolerable fore-pressure P_f of DP1 is 38 Pa. According to the data of Fig. 4.23, in which the slow response of the Pirani gauge was ignored, the chamber on the single DP system must be evacuated in roughing down to 17 Pa (P_s) in order not to raise P_f above P_t just after switching over to fine pumping. On the other hand, on the

Fig. 4.23 Maximum
fore-pressures P_f indicated by
a Pirani gauge just after
switching over the evacuation
mode, as a function of the
switching pressure P_s. (*a*) On
the SEM with the single DP
system and (*b*) with the DP
in-series system [12]

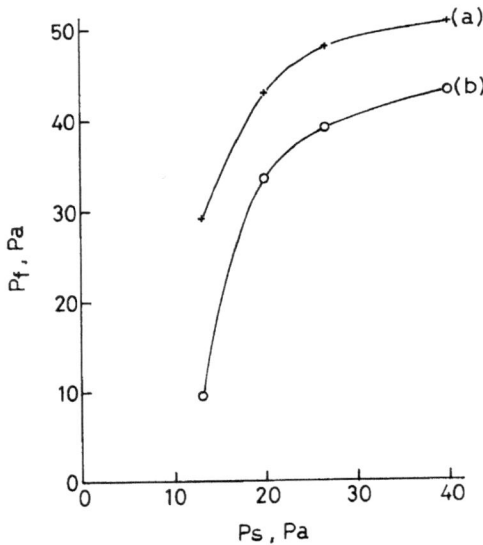

DP in-series system roughing to 24 Pa (P_s) is sufficient enough not to raise P_f above
P_t. Practically speaking, for the SEM chamber of 42 L the roughing time down to
24 Pa was 20 % shorter than the time to 17 Pa. Also, switching the evacuation
modes at a higher pressure is desirable in view of the backstreaming of RP oil vapor.

The Name "Cascade" DP System

I think the readers are unfamiliar with the term "cascade". The cascade DP system
is similar to the DP in-series system, but they are a little bit different with each other.

The word "cascade" in verb means "flow" like a water fall. In the cascade DP
system two gas-flows from the upper pump and from the viewing chamber are gathered by the lower DP, flowing downward, toward the backing RP, which is analogous to a "waterfall".

4.1.8   Advantages of a Small Bypass Valve

I presented a review on "Advantages of slow high-vacuum pumping" in Japanese
2009 [13], where the function of the bypass valve is described in details.

In dynamic DP systems, where the chamber is frequently evacuated from atmospheric pressure to high vacuum, the oil-vapor backstreaming problem sometimes
occurs owing to the excessive gas load flowing into the high-vacuum pump just after
switching the evacuation mode from low-speed roughing to high-speed, high-
vacuum pumping. Two kinds of excessive gas loads exist just after crossover: (1)

gas molecules in the vacuum chamber space and (2) temporarily increased outgassing from the chamber wall surface. The outgassing rate from the chamber wall surface becomes very large with the rapid reduction of pressure owing to high-speed, high-vacuum pumping, because the time constant of diffusion of gas molecules in the wall surface is much larger compared with the time constants of pumping down and the resultant reduction of sorption rate of impinging gas molecules. Slow high-vacuum evacuation, followed by high-speed high-vacuum pumping, is very effective to suppress the temporarily increased outgassing load and the adverse effect of the space gas load and to meet the maximum throughput capacity of the diffusion pump. Equipping with a low-conductance bypass valve makes it possible to use a small-volume buffer tank and a low-speed rotary pump as a backing pump, leading to a reduction in the cost of high-vacuum evacuation systems.

Overload Just After Crossover in Evacuation Systems [13]

Suppose the DP system provided with a bypass valve (3 L/s) as shown in Fig. 4.24. The chamber (~100 L) is first evacuated to a switching pressure Ps (10 Pa) by a roughing RP) (2 L/s), then evacuated through a bypass valve (3 L/s) for about 1 min, followed by high-speed DP pumping (300 L/s). The comparatively low pumping speed of the roughing RP is further reduced when the pressure goes down to the range of 10 Pa and much reduced when approaching the switching pressure of 10 Pa.

By the help of the low-conductance bypass valve, the outgassing rate of the chamber-wall surface could be controlled not to exceed the rate just before crossover, as predicted in Fig. 4.24. Furthermore, the maximum pressure at the inlet port of DP could be suppressed below 0.1 Pa just after crossover because the ratio of bypass conductance (3 L/s) to DP pumping speed (300 L/s) is so small as 0.01.

Slow evacuation through a bypass valve is very effective to suppress the temporarily increased outgassing load and the adverse effect of the space gas load, and to meet the golden rule. (Golden rule: Hablanian uses this term, "golden rule", as follows. The overload due to the outgassing rate can only be prevented by following the golden rule of mass flow limitation.) [14]

Slow evacuation resolves four restrictive conditions for DP systems; (1) roughing RP oil-vapor backstreaming, (2) DP oil-vapor backstreaming, (3) critical backing-pressure problem, and (4) working fluid flow-out. Using a low-conductance bypass valve for a short time makes it possible to use a small-volume buffer tank and a low-speed backing RP, leading to cost reduction of the evacuation system.

Next, suppose the DP system not equipped with a bypass valve. The pressure in the chamber goes down rapidly to high-vacuum after crossover as shown by a broken line in Fig. 4.24. The outgassing rate of the chamber-wall surface becomes very large as shown by the predicted pressure-rise curves drawn by broken lines, because the time constant of diffusing gas molecules in oxide layers is much larger compared to the time constant of the reduction of pressure and that of the reduction of sorption rate [13].

Fig. 4.24 DP system equipped with a bypass valve and the supposed pumping-down and pressure-rise curves in isolation tests. The supposed pumping-down curves and the supposed pressure-rise curves for the conventional DP system, not equipped with a bypass valve, are drawn with *broken lines*. The followings are supposed: Conductance of bypass valve; 3 L/s, chamber volume; 100 L, rated pumping speed of RP; 2 L/s, rated speed of DP system; 300 L/s, switching pressure; 10 Pa. Supposed bypass pumping period: 1 min. ① supposed pressure-rise curves at crossover, ② at the end of bypass pumping, and ③ at an early period of high-speed pumping [13]

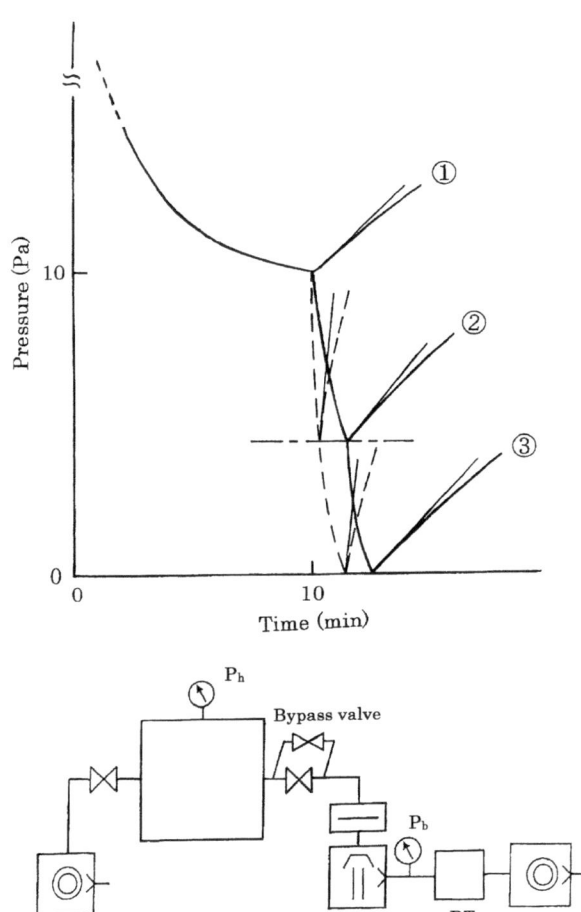

Low-Conductance Bypass Valve, Applicable to the TMP or CP Systems

The pumping performance of the TMP (turbo-molecular pump) is similar to that of the DP. Therefore, a low-conductance bypass valve is applicable to the TMP evacuation system.

A low-conductance bypass valve is also applicable to the CP (cryogenic pump) evacuation system in order to reduce the excess gas load when switching the evacuation mode from RP roughing to CP pumping.

4.2   Analysis and Evaluation on the Vacuum Systems of JEMs

4.2.1   Experiments on Contamination Build-Up

In the JEM with the refined cascade DP system the vicinity of the specimen was effectively evacuated by the radial evacuation system together with the double-cylinder-type anti-contamination device (ACD). We did experiments on contamination build-up in JEM-200CX with the radial evacuation system together with the double-cylinder-type ACD in order to clarify the transfer mechanism of contaminating molecules and the dominant source of them. Experimental results showed that a very clean vacuum is obtained in the vicinity of the specimen [15].

The radial evacuation system with type-1 (double-cylinder type) ACD is presented in Fig. 4.25.

Three types of ACDs were used to control the hydrocarbon partial pressures around the specimen. The ACDs consisted of fins with different shapes, as illustrated in Fig. 4.26. The fins can be cooled down to about $-120\,^\circ\mathrm{C}$ by liq. N_2.

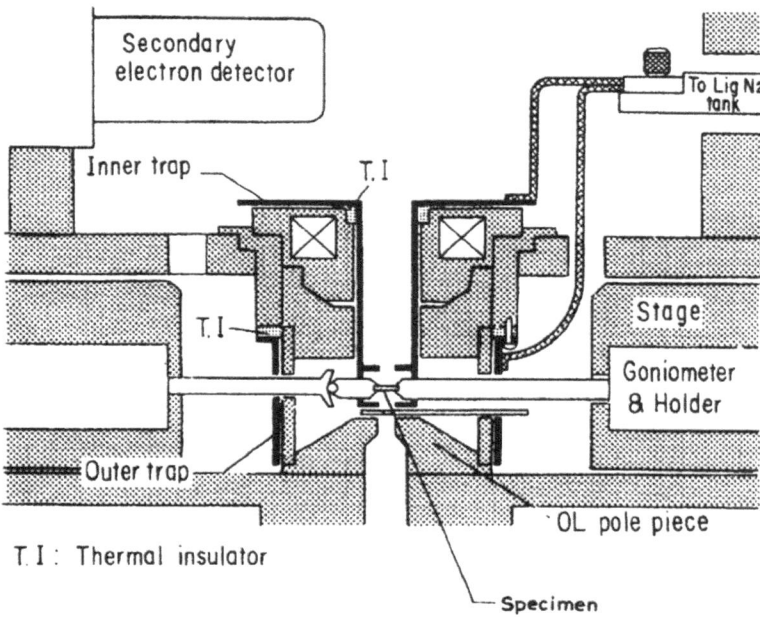

Fig. 4.25 Radial evacuation system with type-1 ACD [15]. T.I.: Thermal insulator

Fig. 4.26 Three types of
ACD with different fins [15]

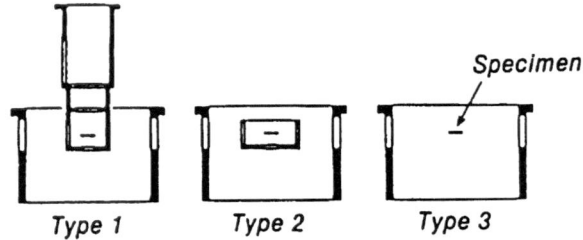

Type 1   Type 2   Type 3

Experiments and Results [15]

To clarify the transfer mechanism of contaminating molecules and the dominant source of them, carbon thin films were irradiated with a fine electron probe of 12 nm diameter, a current density of about 7.6×10^2 A/cm^2 and an accelerating voltage of 200 kV, using three types of ACDs, individually.

The specimen and the specimen cartridge were first immersed in Freon 113 ($C_2F_3Cl_3$) liquid for about 1 h, to chemically eliminate the hydrocarbon molecules adsorbed on their surfaces. The pretreated specimen was then placed in the specimen chamber. Furthermore, the specimen was pretreated with an electron-beam shower (EBS) for 5 min to obtain a clean surface before the irradiation experiment. In the EBS treatment, the beam diameter is about 2 mm, current density about 2×10^{-5} A/cm^2 and accelerating voltage 200 kV.

Two minutes after the EBS treatment, each of 25 spots on the specimen was sequentially irradiated with a fine electron probe for a time of 20 s without using an ACD. The same irradiation process was performed with different elapsed time, 20 and 80 min. Following each series of irradiation processes, secondary-electron images (SEIs) of the irradiated areas were observed with the specimen surface inclined 45° to the electron beam, as presented in Fig. 4.27a, c.

Next, experiments were done using the type-1 ACD. Each of 25 spots on the specimen was irradiated sequentially for 50 s, a few minutes after the EBS treatment. The SEI of the irradiated area showed that contamination deposits did not form on the irradiated areas, but darkening of SEIs on the irradiated areas was observed. A similar process of irradiation was performed with the type-1 ACD for 200 s, at the elapsed times of 0.5, 1.0, 1.5, 2.0, 2.5 and 3 h after the EBS treatment. Contamination deposit build-up induced by a fine electron probe was prevented for 2.0 h after the EBS treatment. Small contamination deposits built up 2.5 and 3.0 h after the EBS treatment. The contamination deposit was larger for longer elapsed times. Two experiments were done in a similar manner with the types 2 and 3 ACD, individually.

The growth rate of a conical deposit was estimated in mL/min for some typical deposits observed without and with types 1, 2 and 3 ACD, individually. The CRs as a function of the elapsed time T after the EBS treatment are presented in Fig. 4.28. In each curve, the growth starts after a certain incubation time (T_0), zero contamination period, and its CR increases with the elapsed time T until it reaches each saturation rate.

Fig. 4.27 SEIs of the irradiated area with the specimen inclined 45° to the electron beam, in three series. In each series, the irradiation started from the *extreme right spot* in the *first row. Small conical images* show the contamination deposits which are arranged along horizontal, parallel lines. *Large black images* show the holes arranged irregularly. The times of 2, 20 and 80 min at the *right side* of each figure show the elapsed time after the EBS treatment for each series. The fine-probe irradiation was conducted without an ACD [15]

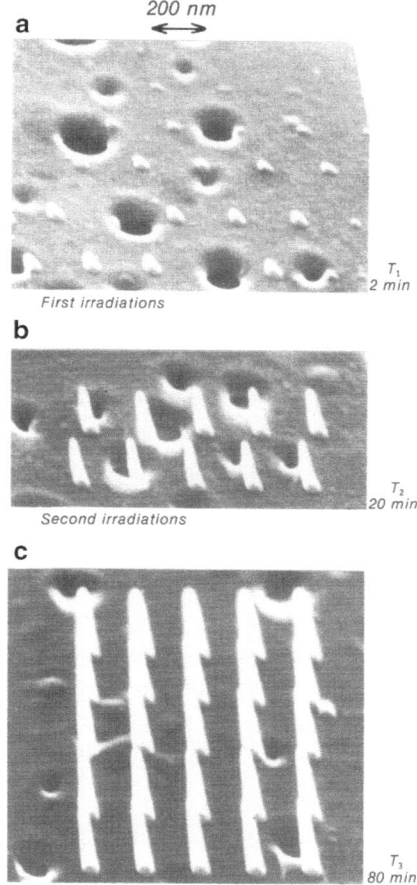

Fig. 4.28 Contamination rates (*CR*s) as a function of the elapsed time *T* after the EBS treatment, obtained without and with types 1, 2 and 3 ACDs, individually [15]

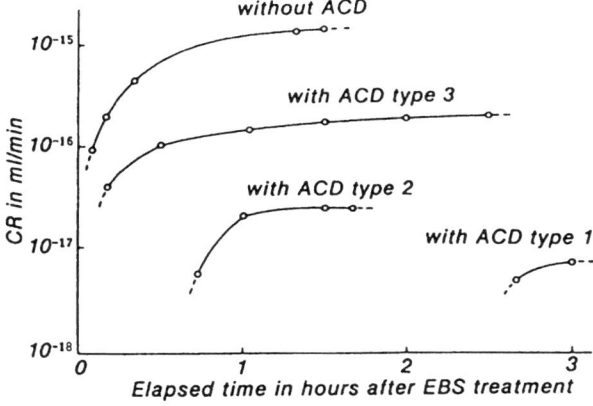

The experimental results of Fig. 4.28 show the following characteristics:

1. The incubation time T_0 exists when an ACD is used. Each CR increases with elapsed time and reaches its saturated rate.
2. The time T_0 and the saturated CR depend largely upon the type of ACD used. The time T_0 is much longer and the saturated CR is much lower with the type-1 ACD than with other types of ACD.

First of all, if the direct condensation to the irradiated area is the dominant transfer mechanism, the CR should be independent of the elapsed time after the EBS treatment, and so the time T_0 should not be observed. Next, if contaminating molecules come from the relatively dirty parts in contact with the specimen by surface diffusion, as Wall considered [16], the time T_0 and the saturated CR should be independent of the type of ACD. The surface diffusion following the adsorption of gases upon the whole specimen surface can explain our experimental results. A clean specimen surface is hit by residual hydrocarbon gases even under an ultrahigh vacuum. Some of the impinging molecules remain on the surface as adsorbates. The amount of adsorbed molecules is known to increase to the saturated amount with the holding time in vacuum. Contaminating molecules will be supplied to the sink by surface diffusion of adsorbed molecules on the area surrounding the sink. However, in an elapsed time shorter than T_0 after the EBS treatment the supply rate will be too low to build up a detectable deposit. After the time T_0, CR will increase with the increase of the amount of adsorbed molecules. Moreover, T_0 will increase with the decrease of residual gas pressure in the vicinity of the specimen, depending on the amount of adsorbed molecules, which is smaller under lower pressure.

When the saturated amount of adsorbed molecules is predominantly governed by equilibrium between adsorption and desorption of molecules, CR (height/time) $\propto 2\pi r_0 / \pi r^2 \propto 1 / r_0$, because the rate of diffusing molecules into the sink is proportional to the circumference of the sink, $2\pi r_0$. On the other hand, when a certain fraction of the hydrocarbon molecules impinging on the whole specimen surface is polymerized in a sink of radius r_0 regardless of any mechanism accepted, CR (height/time) $\propto K / \pi r_0^2 \propto 1 / r_0^2$ (K: constant). In this case the saturated amount of adsorbed molecules and the concentration gradient of them on the surface are predominantly governed by the sink radius r_0. Both cases discussed above are considered to be the extreme cases. In reality, CR is considered to be proportional to $1 / r_0$ $- 1 / r_0^2$.

It can be concluded that in clean ultrahigh vacuum, the model of surface diffusion of adsorbed molecules on the whole specimen surface is most acceptable for the presented experimental results.

Darkening in SEM Images [15]

Using the type-1 ACD, three spots were sequentially irradiated for 1, 5 and 10 min, respectively, directly after the EBS treatment. An SEI of the irradiated area is shown in Fig. 4.29. The darkening images are seen at the irradiated spots and the darkening

Fig. 4.29 Three darkening images after different irradiation times, 1, 5 and 10 min. The fine-probe irradiation was conducted using the type-1 ACD [15]

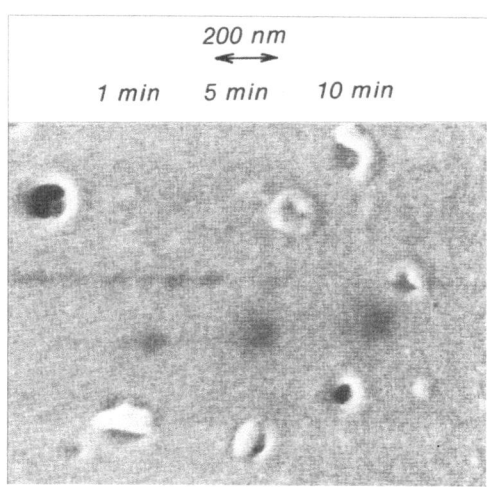

image area seems to enlarge with the irradiation time, while the darkness of images scarcely changes. The darkening line above the three darkening spots is the footprint of a manual spot scan.

The same area of Fig. 4.29 was observed in TEI (transmission electron image) and in STEI (scanning transmission electron image). Their images showed no discernible changes, compared with the virgin area's images before fine-probe irradiation. Therefore, it is reasonable to consider that this darkening observed in SEI shows a decrease in secondary-electron emission at the irradiated spots but not a change in mass of the specimen itself.

The darkening images at the irradiated spots gradually became less dark, and then were no longer visible about 3 h after fine-probe irradiation [15].

It could be said that JEMs provided with the refined cascade DP system, with the radial evacuation system with the double-cylinder type (type-1) ACD achieved a clean vacuum in the vicinity of the specimen.

4.2.2   Pressure Simulation by the Resistor-Network Simulator

The pressure in the vicinity of the specimen in an EM cannot be measured by a commercial vacuum gauge. Under such actual situation we must estimate the pressures around the specimen, positioned in the objective lens pole-pieces.

Basic concepts of outgassing sources and high-vacuum pumps and their characteristic values have been briefly introduced in Sect. 3.1.1. Based on this basic concept one can design the vacuum circuit for the complex high vacuum system, and so the pressures at every position can be simulated [17, 18].

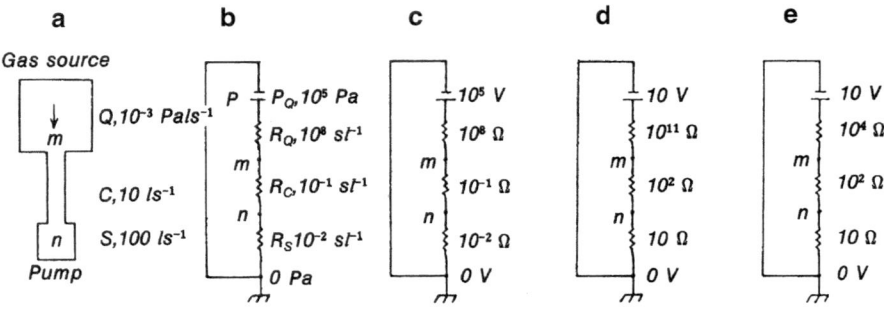

Fig. 4.30 Steps for designing a simulator circuit of a vacuum system [17]

Procedure for Designing Resistor-Network Simulator Circuit [15]

The typical procedure for simulating the simplest system is shown in (a) to (e) of Fig. 4.30.

(a) Let us estimate the outgassing rate Q, conductance C, and pumping speed S of the system as 10^{-3} Pa L/s, 10 L/s, and 100 L/s, respectively.

(b) Draw the corresponding vacuum circuit (b). Note that the value P_Q is set to 10^5 Pa which is large enough as compared with the vacuum pressure.

(c) Convert the vacuum circuit into the corresponding electric circuit (c), that is, Pa into V (volts) and s/L into Ω (Ohms).

   In this circuit, the voltage generator of 10^5 V corresponds to the pressure generator of 10^5 Pa, and the electric resistors correspond to the vacuum resistors.

   In the circuit (c) 10^5 V of the voltage generator is too large and 10^{-2} and 10^{-1} Ω of the resistors are too small to assemble a practical simulator. Therefore, the following modifications are made as shown in (d) and (e).

(d) (i) Multiply the voltage of the generator 10^5 V by 10^{-4}. As a result, the reconverting factor k_1 from voltage V to pressure Pa becomes 10^4 Pa/V.

   (ii) Multiply the resistances of all resistors in circuit (c) by 10^3.

       It should be noted that the relative distribution of the voltage in the circuit (d) is not changed by these modifications. In this case the reconverting factor k_1 remains 10^4 Pa/V. In this circuit, however, 10^{11} Ω for the resistor is too large to assemble a simulator.

(e) Further modify the circuit into the simulator circuit (e) in which 10^4 Ω is used in place of 10^{11} Ω. The resistor of 10^4 Ω is easily obtainable.

Note that 10^4 Ω is still large enough compared with other resistances, 10^2 and 10 Ω. So, the relative distribution of voltage in the circuit is kept constant with an error of 1 %.

   As a result, the reconverting factor k_2 from V to Pa in (e) is given by $k_2 = 10^{-7} \times k_1 = 10^{-3}$ Pa/V.

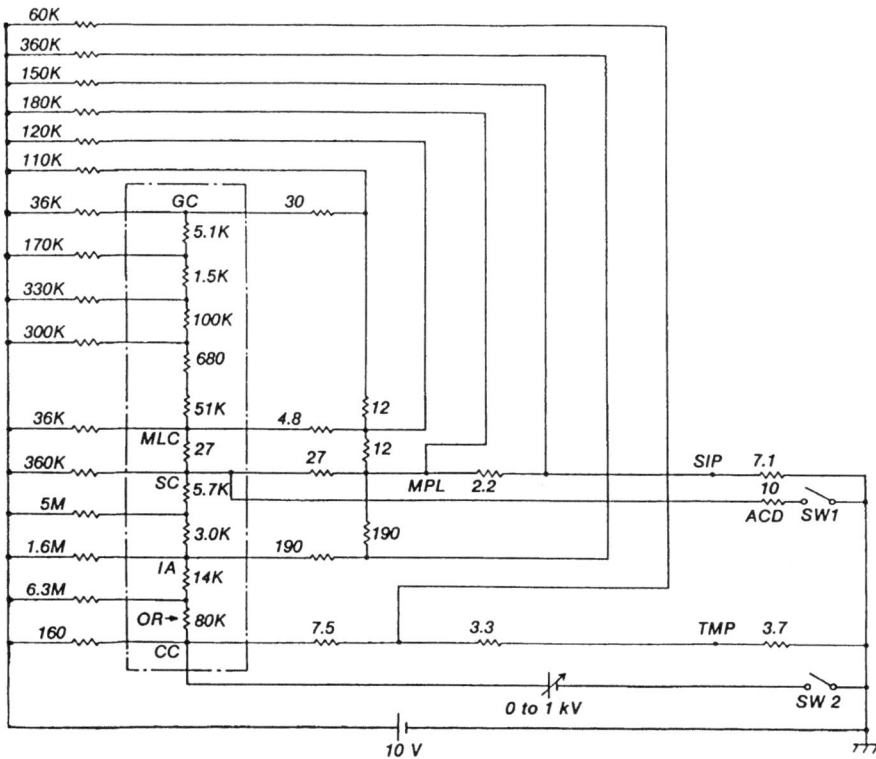

Fig. 4.31 Circuit of the simulator for the high-vacuum system of an EM. The outlined part corresponds to the whole column. The resistors are represented in Ω and the symbols K and M used for resistance mean $\times 10^3$ and $\times 10^6$, respectively [17]

The procedure for designing the simulator circuit of the high-vacuum system of an electron microscope is the same as the one described above, although the actual high-vacuum system has many elements. Figure 4.31 shows the simulator circuit of the high-vacuum system of an EM, where the reconverting factor is 10^{-3} Pa/V.

In the simulator circuit of Fig. 4.31, the outlined part corresponds to the whole column. The resistors along the center line between the gun chamber (GC) and the camera chamber (CC) correspond to the resistance of an orifice, conducting pipes, apertures, etc. The resistors located on the left side of the outlined part correspond to the outgassing rates of each part of the whole column. The resistors located above the outlined part correspond to the outgassing rates of the evacuation pipes connecting the pumps to the column. These resistors related to the outgassing rates are to be connected to the voltage generator which supplies 10 V as a whole.

The 10 Ω resistor connected to switch 1 (SW1) corresponds to the cryo-pumping speed of an anti-contamination device (ACD) consisting of liq. N_2-cooled fins. The pumping speed of this device is estimated to be 100 L/s. The voltage generator with

Fig. 4.32 Pressure
distribution in the column [17]

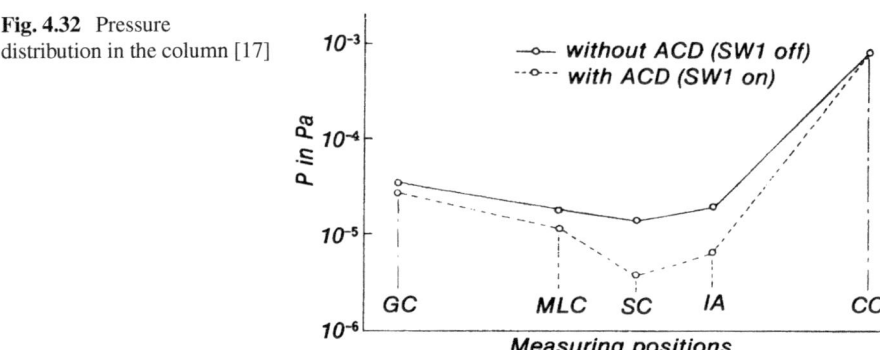

a variable voltage range of 0 to 1 kV, connected to CC (camera chamber) through SW2, corresponds to variable pressures of 0 to 1 Pa.

Figure 4.32 shows the simulated pressures at some positions in the microscope column. The pressure-distribution curves indicate the minimum pressure at the specimen chamber SC. This proves that the ACD is effective for the specimen chamber.

Hirano and Kondo et al. [19] described an algorithm for matrix analysis using a micro-computer for analyzing vacuum circuits. The matrix-analysis method was applied to two high-vacuum systems, an outgassing pipe and an EM.

Today, you can use the commercial electronic circuit simulation software "Spice", in order to simulate the pressures of every position of the complex high vacuum systems, based on the step (b) of Fig. 4.30.

Kendall [20] evaluate our article [17] by submitting Comments on: "Resistor-network simulation method for a vacuum system in a molecular flow region" [20].

References

1. Holland L (1971) Vacua: how they may be improved or impaired by vacuum pumps and traps. Vacuum 21(1/2):45–53
2. Yoshimura N, Hirano H, Norioka S, Etoh T (1984) A cascade diffusion pump system for an electron microscope. J Vac Sci Technol A 2(1):61–67
3. Hengevoss J, Huber WK (1963) The influence of fore-vacuum conditions on ultrahigh-vacuum pumping systems with oil diffusion pumps. Vacuum 13(1):1–9
4. Kojima, Nagahama Y, Yoshimura N, Oikawa H (1973) Improved vacuum system for the electron microscope. J Vac Soc Jpn 16(6):217–224 (in Japanese)
5. Mikami O, Oikawa H, Yoshimura N, JEOL Co Ltd (1969) Patent no 1969–31,075, 22 Dec 1969
6. Yoshimura N, Ohmori S, Nagahama Y, Oikawa H (1974) A new vacuum system for an electron microscope. In: Proceedings of the 6th international vacuum congress. J J Appl Phys Suppl 2 Pt 1: 249–252
7. Laurenson L (1982) Technology and applications of pumping fluids. J Vac Sci Technol 20(4):989–995

8. Nakayama K (1965) Vapor pressure of vacuum pump oils. J Vac Soc Jpn 8(10):333–337 (in Japanese)
9. Hablanian MH, Maliakal JC (1973) Advances in diffusion pump technology. J Vac Sci Technol 10(1):58–64
10. Hirano H, Yoshimura N (1981) Thermal loss of a cold trap. J Vac Soc Jpn 24(4):167–169 (in Japanese)
11. Yoshimura N, Kobayashi N, JEOL Co Ltd (1984) Patent no 1984–17,083, 18 May 1984
12. Norioka S, Yoshimura N (1991) Practical advantages of a cascade diffusion pump system of a scanning electron microscope. J Vac Sci Technol A 9(4):2384–2388
13. Yoshimura N (2009) Advantages of slow high-vacuum pumping for suppressing excessive gas load in dynamic evacuation systems. J Vac Soc Jpn 52(2):92–98 (in Japanese)
14. Hablanian MH (1992) Prevention of overload in high vacuum systems. J. Vac. Sci. Technol. A 10(4):2629–2632
15. Yoshimura N, Hirano H, Etoh T (1983) Mechanism of contamination build-up induced by fine electron probe irradiation. Vacuum 33(7):391–395
16. Wall JS (1980) Contamination in the SEM at Ultrahigh Vacuum Scan. Electron Microsc 1:99–106.
17. Ohta S, Yoshimura N, Hirano H (1983) Resistor-network simulation method for a vacuum system in a molecular flow region. J Vac Sci Technol A 1(1):84–89
18. Yoshimura N (1985) A differential pressure-rise method for measuring the net outgassing rates of a solid material and for estimating its characteristic values as a gas source. J Vac Sci Technol A 3(6):2177–2183
19. Hirano H, Kondo Y, Yoshimura N (1988) Matrix calculation of pressures in high-vacuum systems. J Vac Sci Technol A 6(5):2865–2869
20. Kendall BRF (1983) Comments on: resistor-network simulation method for a vacuum system in a molecular flow region. J Vac Sci Technol A 1(84); J Vac Sci Technol A 1(4): 1881–1882

Chapter 5
Development of JEOL SIPs

Abstract Philips (Holland) put an EM equipped with a large SIP place on the market early 1980s first in the world. The SIP had three ports for the gun chamber, for the specimen chamber, and for the evacuation pipe to the diffusion pump. The SIP was apparently dedicated for the Philips EMs. We had to develop the JEOL SIPs dedicated for our JEMs, as Philips did. We must develop two types of SIP, one having high pumping speeds in ultrahigh-vacuum region, and the other having stable pumping performance for inert gases such as Ar for thinning specimens in AES (Auger-electron spectrometer), together with the good performance in UHV.

5.1 Papers, Which Helped Us Much in Developing the JEOL SIPs

We studied the Penning discharge itself, the basis of the pumping mechanism of SIP. We got the knowledge of SIP much from the following five papers, by Jepsen [1], by Rutherford [2], by Andrew [3], by Jepsen et al. [4], and by Baker and Laurenson [5].

5.1.1 "The Physics of Sputter-Ion Pumps" by Jepsen [1]

Jepsen presented the basis of SIPs (1968) [1], in which "Discharge intensity", "Pumping mechanisms", "The 'energetic-neutrals' hypothesis" are described in details. Section "Discharge intensity" is the basis of the performance of SIP in UHV region, and Section "The 'energetic-neutrals' hypothesis" describes the behavior of Varian triode pump for inert gas pumping.

N. Yoshimura, *Historical Evolution Toward Achieving Ultrahigh Vacuum in JEOL Electron Microscopes*, SpringerBriefs in Applied Sciences and Technology, DOI 10.1007/978-4-431-54448-7_5, © The Author(s) 2014

Discharge Intensity [1]

To be of utility in applications requiring a high pumping speed, it is important to achieve high discharge intensities (I/P). There are many different combinations of electrode geometry, voltage and magnetic field which will yield a given value of I/P at a particular pressure and for a particular species of gas. For many applications it is important that high values of I/P be maintained over a wide pressure range. Since the variation of I/P with pressure also depends strongly on electrode geometry, voltage and magnetic field, the allowable range of these parameters is further restricted. Additional factors which must be taken into account in designing pumps include gas conductance into the discharge region, pump heating during high-pressure operation, physical size and shape of the pump, weight of the pump and manufacturing cost. The diversity of pump designs that has emerged over the past several years is not surprising in view of the options available to the pump designer, both in terms of allowable ranges of the basic parameters and in terms of the specific design objectives and constraints.

Useful insights into the dependences of discharge intensity on the relevant parameters can be obtained through certain elementary, and quite approximate, considerations involving space-charge depression of potential [6]. Let us consider first a 'long-anode' Penning cell operating at pressures sufficiently low that the number of positive ions present in the discharge is much less than the number of electrons. (For parameters in the range commonly used in SIPs, this implies a pressure below about 10^{-5} Torr.) In this case we can neglect positive-ion neutralization of the electron space charge.

Even though the actual distribution of electrons in the discharge is in general far from uniform, it is instructive to calculate the depression of potential produced by a uniformly distributed space charge, and also to compute the maximum amount of charge (again uniformly distributed) that can be 'contained' in the discharge cell. To first order, at least, the discharge intensity should be proportional to the number of electrons present.

Starting with the divergence equation (MKS units),

$$\nabla \cdot E = \rho / \varepsilon_0 ,$$

we can show that for $L \gg r_a$ (L being the anode length and r_a the anode radius) the radial electric field E_r inside the anode is

$$E_r = (\rho / 2\varepsilon_0)r$$

($\rho < 0$ because the space charge is due to electrons.) Neglecting end effects, the potential V_0 on the cell axis would be equal to the anode potential V_a in the absence of a space charge. For a total charge q contained within the cell,

$$V_0 = V_a + q / 4\pi\varepsilon_0 L. \quad (q<0).$$

The maximum charge q_{mp} that can be contained within the anode is that for which V_0 is reduced to the cathode potential ($V_0 = 0$). Thus

$$q_{mp} = -4\pi\varepsilon_0 V_a L. \tag{5.1}$$

As one might expect, q_{mp} is linearly proportional to both V_a and L. Somewhat surprisingly, however, q_{mp} is independent of anode diameter. This accounts for the fact that high-speed SIPs employ many small-diameter cells in parallel rather than just a few cells of relatively large diameter. Also, the significance of the dependence of q_{mp} on the anode voltage has not been lost on the designers of SIPs.

In view of the independence of q_{mp} of the anode diameter, why are not cells of much smaller diameter employed in SIPs? One reason is that the magnetic field must be increased as the cell diameter is reduced. An elementary scaling relation is

$$Br_a = \text{const.}$$

for constant discharge intensity, with the other relevant parameters held constant [6].

Similar calculations on the potential depression and maximum space charge have been performed for both 'normal' and 'inverted' magnetrons. Let r_o be the radius of the outer cylinder, and r_i be the radius of the inner cylinder. It is found that

$$q_{mn} = -\frac{4\pi\varepsilon_0 V_a L}{1 - \left\{2r_i^2 \left(r_o^2 - r_i^2\right)\right\} \ln\left(r_o / r_i\right)}$$

for the normal magnetron, and

$$q_{mi} = -\frac{4\pi\varepsilon_0 V_a L}{-1 + \left\{2r_o^2 \left(r_o^2 - r_i^2\right)\right\} \ln\left(r_o / r_i\right)}$$

for the inverted magnetron. In the filamentary cathode limit for the normal magnetron,

$$q_{mn}\Big|_{r_i \ll r_o} \cong -4\pi\varepsilon_0 V_a L = q_{mp} \, q_{mn},$$

and in the filamentary anode limit for the inverted magnetron,

$$q_{mi}\Big|_{r_i \ll r_o} \cong -\frac{4\beta_0 V_a L}{2 \ln\left(r_o / r_i\right)} = \frac{q_{mp}}{2 \ln\left(r_o / r_i\right)}.$$

It is worth noting that the maximum charge that can be contained in a Penning cell is the *same* as in a normal magnetron employing a filamentary cathode. This is not surprising, since the electron space charge in a Penning cell converts it from a 'parallel-field' device to a 'crossed-field' device of the normal magnetron variety. The inverted magnetron with filamentary anode, however, is clearly inferior in its charge-containing ability, and should thus be less useful in SIPs than normal magnetrons and Penning cells.

An order-of-magnitude estimate of discharge intensity may be made as follows:

$$I = -q / \overline{\tau}_c,$$

or

$$I / P = -q / \overline{\tau}_c P$$

where $\overline{\tau}_c$ is an average time between ionizing collisions for all electrons in the discharge. For $V_a = 7 \times 10^3 \, \text{V}$ and $L = 2.5 \times 10^{-2} \, \text{m}$, $q_{mp} \cong -2 \times 10^{-8} \, \text{C}$ (from Eq. (5.1)). If all of the electrons have at all times near-optimum energy for producing ionizing collisions, then $\overline{\tau}_c P \cong 1.5 \times 10^{10}$ Torr•s for gases such as argon and nitrogen. Thus

$$\left(\frac{I}{P} \right)_{mp} = -\frac{q_{mp}}{\overline{\tau}_c P} \cong \frac{2 \times 10^{-8} C}{1.5 \times 10^{-10} Torr \cdot s} \cong 130 \text{ A / Torr.}$$

This value of I/P is unrealistically high, since the electrons do not have near-optimum energy at all times. This is to be compared with I/P values of 20–60 A/Torr found experimentally in the 10^{-8}–10^{-5} Torr pressure range. In view of the crudeness of the model and the approximations employed, this agreement between experiment and theory may be regarded as eminently satisfactory.

One of the most extensively studied—and yet least understood—facets of magnetically confined cold-cathode gas discharges is their pressure dependence. No attempt will be made to review this important topic here. Instead we shall proceed to a discussion of some of the mechanisms by which pumping of gas occurs in SIPs.

The 'Energetic-Neutrals' Hypothesis [1]

One of the explanations advanced has been that argon ions are transformed into neutral atoms of essentially the same kinetic energy through charge-exchange collisions with argon atoms in the gas phase. Since energetic neutrals are unaffected by electric fields, they are capable of reaching such surfaces as the anode or the pump envelope. Such an explanation would predict a pumping speed which is proportional to pressure for a constant discharge intensity (I/P). Experimental measurements, however, show quite conclusively that the pumping speed varies no more rapidly with pressure than does I/P. An alternative explanation must therefore be sought.

The explanation proposed here is that energetic neutral argon atoms are indeed responsible for argon pumping; but the dominant mechanism for the production of these energetic neutrals is through impact of argon ions at the cathodes rather than through charge exchange in gas-phase collisions.*(*Although the discussion in this paper centers around argon, the concepts involved are applicable to the other noble gases, both heavier and lighter.)

To account quantitatively for the observed pumping of argon in triode pumps, one atom of argon must be pumped for about every 10 argon ions generated in the discharge. If the energetic neutrals have a sticking probability of, say, 0.5, then about 20 % of the argon ions incident upon the cathodes must result in the emission of such energetic neutral argon atoms.

Table 5.1 E_1 / E_0 for various scattering angles (θ_1) and target materials, with argon as the incident ion [1]

Target material	E_1 / E_0				
	$\theta_1 = 0$	$\theta_1 = \pi / 6$	$\theta_1 = \pi / 4$	$\theta_1 = \pi / 2$	$\theta_1 = \pi$
Ti	1.00	0.74	0.58	0.09	0.01
Mo	1.00	0.84	0.72	0.37	0.14
Ta	1.00	0.94	0.88	0.64	0.41

Let us now examine the plausibility of this explanation. A growing body of evidence—of which the present work is a part—strongly suggests that for ion energies up to about 10 keV a large fraction of the ions incident upon a metal surface engage in elastic collisions with the first two or three atom layers, that most are neutralized as they approach the metal surface or during the collision process, and that most of those which escape do so as neutrals.

Perhaps the work most directly relevant to the problem at hand is that reported by Kornelsen [7]. He found that the sticking probability for argon ions incident upon a tungsten target reaches a 'plateau' value of about 0.60 in the energy range ~1–5 keV. By assuming that scattering is isotropic in the center-of-mass system, he was able to show by simple calculation that indeed about 40 % of the incident ions should be reflected as neutral atoms. Support for the validity of Kornelsen's results is contained in work by Winters and Kay [8].

It is instructive to compute the energy with which an incident ion of mass m_1 is scattered upon making a single elastic collision with a target atom of m_2. From conservation of energy and momentum, it is readily shown that

$$\frac{E_1}{E_0} = \frac{1}{\left(1 + m_2 / m_1\right)^2} \left[\cos \theta_1 \pm \sqrt{\left\{\left(m_2 / m_1\right)^2 - \sin^2 \theta_1\right\}} \right]^2 , \qquad (5.2)$$

where θ_1 is the scattering angle for the incident particle, E_0 is the kinetic energy of the incident particle and E_1 is the kinetic energy of the incident particle after scattering. Equation (5.2) clearly shows that if $m_2 < m_1$, scattering can occur only in the forward direction $\left(\theta_1 < \pi / 2\right)$. For $m_1 < m_2$, all scattering angles are allowed.

For normal incidence, it is necessary that $\pi / 2 \leq \theta_1 \leq 3\pi / 2$ for the scattered particle to escape from the surface. For grazing incidence, however, escape occurs for $0 \leq \theta_1 \leq \pi$.

Table 5.1 shows E_1 / E_0 for Ar^+ ions ($M \cong 40$) incident upon Ti ($M \cong 48$), Mo ($M \cong 96$) and Ta ($M \cong 181$) surfaces for selected scattering angles in the range 0–π.

From Kornelsen's work, the sticking probability of argon ions in tungsten is 0.1 at about 250 eV, and reaches a value of 0.5 at about 700 eV. Although we might anticipate that the sticking probabilities would be higher in titanium (because of its lower atomic mass), it seems likely that energies of a few hundred electron volts will be required for sticking probabilities of argon in titanium to exceed 0.5. There is some evidence in the literature that neutral atoms have somewhat lower sticking probabilities than do ions of the same kinetic energy. Despite this, it appears that sticking probabilities sufficiently large to produce significant pumping can exit for neutral-atom kinetic energies of several hundred electron volts and above.

In the case of diode pumps employing flat titanium cathodes, most of the argon ions will strike the cathodes at near-normal incidence with energies in the ~1-5 keV range. Because the atomic mass of titanium is only 20 % greater than that of argon, scattering is predominantly in the forward direction. The fraction back-scattered will be relatively small, and the energies will be less than 10 % of the incident-ion energies, leading to low sticking probabilities. Since the back-scattered atoms can go to the anode, this mechanism may well account for that pumping of argon which is observed to take place there. For helium striking titanium, however, $m_2/m_1 = 12$, so that nearly half of the incident ions could be backscattered as neutrals with energies in excess of about 70 % of the incident-ion energies. This should lead to pumping at the anode, with speed limited by the rate of sputtering from the cathodes, which is low for helium.

In the case of single-voltage triode pumps, we might expect about 50 % of the ions to strike the sides of the cathode strips at near-grazing incidence. Even with titanium cathodes, we might expect 30-40 % of these to be scattered from the surface, predominantly forward and with energies in excess of 30 % of the incident-ion energy. This could easily result in one atom of argon being pumped for about each 10 argon ions generated in the discharge, thereby accounting at least semi-quantitatively for the observed pumping speed for argon.

The cathodes used in Varian's triode pumps employ parallel strips of titanium. In one experiment, alternate strips of titanium and molybdenum were employed. This led to an increase in argon pumping speed by a factor of about 1.5. In another experiment, alternate strips of titanium and tantalum were used. This led to an even greater increase in argon-pumping speed. We might expect that about 50 % of the ions would strike the edges of the cathode strips at near-normal incidence. These should produce few energetic neutrals from Ti, many more from Mo and still more from Ta. This more effective utilization of the normal-incidence ions could account for most of the observed increase in speed.

In 1966 Tom and James [9] presented a paper entitled 'Inert-Gas Ion Pumping Using Differential Sputter-Yield Cathodes.' One cathode was made of titanium and the other of an unidentified material which, assertively, possesses a much higher sputtering rate than does titanium. It is generally believed that the unidentified material is tantalum. Since the sputtering rate for tantalum is almost identical with that for titanium, it is most unlikely that the differential in the sputter yields could account for the observed performance. An alternative explanation involving diffusion of argon into the interior of the tantalum cathode (in analogy with diffusion of helium into titanium) appears to be of dubious validity, as the diffusion coefficient would need to be unexpectedly high.

In the light of our discussion of the role played by energetic neutrals in triode pumps, a much more plausible explanation is that these are responsible for most of the observed pumping of argon in which one or both of the cathodes are tantalum. The significant factor here is that back-scattering of energetic neutrals from tantalum should be much greater, both in numbers and in energy, than from titanium because of tantalum's several-fold-higher atomic mass. If the explanation proposed here is correct, then we would expect that most of the steady-state pumping of argon occurs at the anode.

In a comparison between a standard Varian triode pump (titanium cathodes) and a Varian diode pump with one tantalum and one titanium cathode, argon speeds were found to be about 21 % and about 25 % of the nitrogen speeds for the respective pumps. Using a partial-pressure gauge, currents at the $M=40$ and $M=28$ peaks were measured while pumping against an air leak. Defining $R \equiv I_{40} / I_{28}$, it was found that $R \cong 0.03$ for the triode pump and $R \cong 0.09$ for the Ta-Ti diode pump. In view of the somewhat greater argon speed for the Ta-Ti diode pump, one would expect R to be somewhat smaller than for the triode. Yet it is larger by a factor of about 3. Thus the presence of oxygen and nitrogen appears to reduce the argon-pumping speed for the Ta-Ti diode pump rather significantly. In terms of our energetic-neutrals hypothesis, a possible explanation for this is that back-scattering of argon from tantalum is inhibited by the presence of adsorbed gas (oxygen and nitrogen) on the cathode surfaces when pumping against an air leak. The first atom layer encountered by an incident ion is more important to back-scattering than is the second layer. Because of the low atomic masses of oxygen and nitrogen relative to argon, scattering from them will be in the forward rather than in the backward direction. This effect is less important in a triode pump because of the near-grazing incidence of the ions.

Another type of SIP is the magnetron pump. Here, again, one would expect energetic neutrals to play an important role in the pumping of argon. Cathode diameter should be important in respect to the dependence of scattering on angle of incidence, and cathode material should influence strongly the amount and energy of backscattering. For designs such that energetic neutrals play a dominant role, much of the argon should be pumped at the anode.

5.1.2   *"Sputter-Ion Pumps for Low Pressure Operation"* *by Rutherford [2]*

Rutherford investigated the relationship between I/P (discharge intensity) and B × d (product of magnetic field and anode-cell diameter) [2].

I/P (A/Torr) vs. P (Torr) for SIPs Operated at Three Magnetic Fields [2]

Upon performing a further set of experiments in which magnetic field was varied as a parameter, and plotting discharge intensity vs. pressure, a significant change in cut-off pressure was noted as shown in Fig. 5.1 (I/P vs. P, V = 3 kV, 1/2 in. cells, for a typical range of operating magnetic fields, B = 1,000, 1,500 and 2,000 G.) As magnetic field is increased by a factor of 2, the cut-off pressure is reduced from about 2×10^{-7} Torr to 5×10^{-10} Torr. The sharp dependence of cut-off pressure on magnetic field displayed in Fig. 5.1 is just the kind of phenomenon which could explain the behavioral difference noted above between large and small pumps (larger magnetic fields for larger pumps).

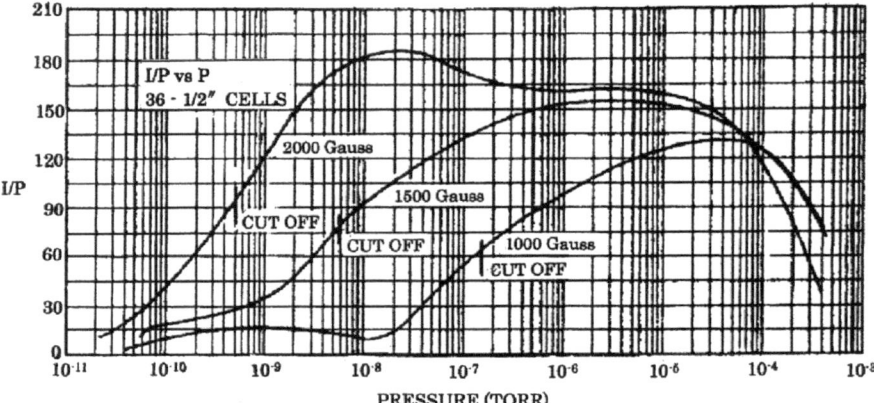

Fig. 5.1 Discharge intensity I/P (A/Torr) vs. pressure P (Torr) for SIP operated at three different magnetic fields. B = 1,000, 1,500, 2,000 G. V = 3 kV, anode cell diameter = 1/2 in. [2]. *Notes*: "Cut-off" pressure is defined as the pressure where discharge intensity has fallen to one-half its value at high pressure (1×10^{-5} Torr). *Note*: 1 G = 10^{-4} T

It can be said from Fig. 5.1 that the larger the magnetic field the better performance in the very low pressure region can be expected.

I/P (A/Torr) vs. P (Torr) for SIPs with Three Different Anode-Cell Diameters [2]

Many effects at high pressures (up to ~10^{-4} Torr) in the Penning discharge are closely related to B×d, the product of magnetic field B and anode-cell diameter d. Typical of this dependence is the minimum operating field for a given cell diameter (B×d ≃ 0.2 kGauss•inch). Because of these dependences on the B×d product, and because the magnetic field affects the cut-off pressure, it was decided to increase the B×d product by changing the anode cell diameter, keeping the magnetic field constant, to determine whether this would also change the cut-off pressure. The result of this variation is shown in Fig. 5.2 (I/P vs. P, for 1/2 in., 1 in., and 2 in. diameter anode cells, V = 3 kV, B = 1,000 G). It can be seen that the cut-off pressure at 1,000 G is about 1×10^{-7} Torr for 1/2 in. diameter anode cells, about 1×10^{-9} Torr for 1 in. diameter anode cells, and below 1×10^{-11} Torr for 2 in. diameter anode cells. In these experiments, the aspect ratio (ratio of anode cell length to anode cell diameter) is kept constant at 1.5, and the number of anode cells varies as follows: 36-1/2-in.-diameter cells, 6–1 in. diameter cells and 3–2 in. diameter cells.

Since the practical magnetic fields obtainable from permanent magnets are limited in magnitude, the results shown in Fig. 5.2 indicate a method for reducing the cut-off pressure in SIPs to very low pressures while using easily available magnetic fields.

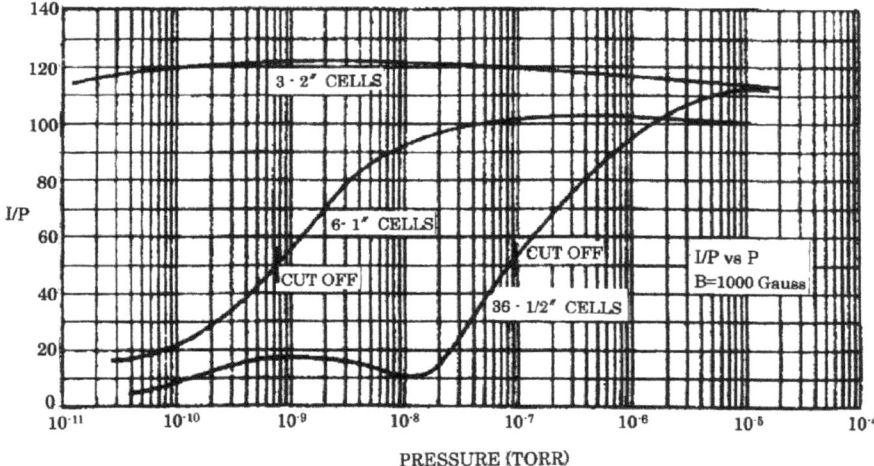

Fig. 5.2 I/P (A/Torr) vs. P (Torr) for SIPs with three different anode cell diameters. d = 1/2, 1 and 2 in., B = 1,000 G, V = 3 kV [2]

Notes

At a glance, 2-in. cell is best for extremely low pressure operation. However, the larger the cell diameter will be, the longer the length when keeping the aspect ratio. And, the number of the anode cell is inversely proportional to the square of cell diameter. In Fig. 5.2 the area for 2-in. cells is 2 times the area for 1-in. cells.

5.1.3 "The Development of Sputter-Ion Pumps" by Andrew [3]

Andrew [3] reviewed pump configurations (presented in Fig. 5.3) which were investigated for improved inert-gas pumping, and discussed their performance.

Abstract [3]

The theory and practice of the Penning discharge, which is the basis of all SIPs, is briefly reviewed. A model for the pumping mechanism in this cell is discussed.

The development of the geometry of the Penning cell to meet the requirements of high pumping speed and low-pressure operation is described, and the methods and reasons for designing various forms of pumping apparatus are discussed. Improvements in pump-cell design aimed at increasing the inert-gas pumping speed relative to the air speed are described.

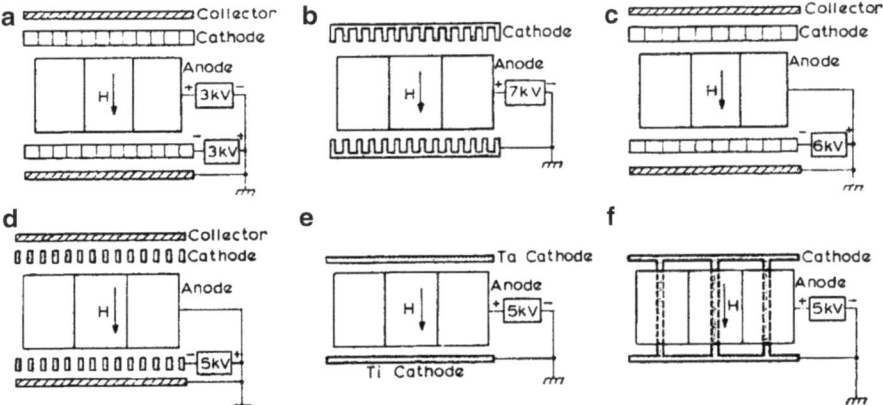

Fig. 5.3 Pump designs for improved inert-gas pumping. (Andrew, 1968) [3]. (**a**) Triode, Brubaker 1959 [10]. (**b**) Slotted cathode diode, Jepsen et al. (1960) [4]. (**c**) Triode, Hamilton (1961) [19]. (**d**) Triode, Varian Noble Ion Pump (1967). (**e**) Differential ion pump, Tom and James (1966) [9]. (**f**) Magnetron pump Andrew et al. (1968) [20]

The problem of unstable pumping of argon is discussed, and an explanation based on mode changes of the discharge is supported by experimental observations.

Future refinements of the SIP involving the use of new cathode materials and cell geometries may give additional benefits in performance, but it is becoming increasingly important to develop a method of evaluating the performance of sputter-ion pumps so that the influence of changes in design can be correctly assessed.

Pump Configurations for Improved Inert-Gas Pumping [3]

Typical pump designs for improved inert-gas pumping is presented in Fig. 5.3.

The triode pump described by Brubaker [10] had cellular cathodes and two collector electrodes (Fig. 5.3a). Ion bombardment of the walls of the cells at oblique incidence resulted in a greater yield of sputtered material which trapped gas at the collector. Ions decelerated before reaching the collector have insufficient energy to cause sputtering and re-emission of the trapped gas.

Jepsen et al. [4] introduced the slotted-cathode diode pump (Fig. 5.3b). The oblique bombardment of the sides of the slots in the cathode, together with the large areas of cathode where trapping could occur, resulted in an increase of the argon speed up to 6 %. It was claimed that this design was simpler and more economic compared to the triode.

Hamilton (1962) showed that by rearranging the potentials on the triode, both the air pumping speed and the argon speed could be improved, the latter rising to 25 %.

This arrangement simplified the power supply and the construction of the triode, and provided an adequate counter argument to that of Jepsen et al. [4]. Hamilton was unable to provide a satisfactory explanation of his results, which appear to be at variance with the model of the pumping mechanism developed to date (Fig. 5.3c).

Varian have manufactured a triode having cathodes with closely spaced strips (Fig. 5.3d) resembling the surface achieved in the slotted-cathode design. Bance and Craig [11] have described a triode with mesh cathodes.

Carter [12] suggested that tantalum, on account of its better sputtering and trapping properties, should have advantages, when compared to titanium as a cathode material. Tom and James [9] used two different materials, believed to be tantalum and titanium, for the cathodes in a standard diode cell (Fig. 5.3e) and found argon speeds up to 25 % of the air speed.

Another approach, using a magnetron structure, has yielded argon speeds between 12 and 20 % of the air speed. The inert-gas pumping process in this pump is the subject of another paper at this Congress [3].

5.1.4 "Stabilized Air Pumping with Diode Type Getter-Ion Pumps" by Jepsen et al. [4]

Jepsen et al. [4] presented the details of the diode pump with slotted cathodes. They also described the "argon problem" and the magnetic field efficiency [4].

The Argon Problem [4]

In most pumping applications there is little need for high pumping speeds for the noble gases. In a limited number of cases the pumping of argon becomes somewhat important because of the approximately 1 % argon content of the earth's atmosphere. In a still more restricted class of applications, high speed pumping of particular noble gases is of primary importance.

Let us now turn our attention to the second situation just mentioned. Under certain conditions of pumping against a continuous air leak, the conventional diode type of getter-ion pump may exhibit periodic pressure fluctuations. Since these fluctuations are associated with the argon which is normally present in air, they have been given the name "argon instability". Some of the characteristics of argon instability are the following:

- To get a new or freshly cleaned pump to go unstable, it is typically necessary to operate it at a pressure of 1×10^{-5} Torr for several hundred hours; at $P = 1 \times 10^{-6}$ Torr, several thousand hours would be required. There is, however, wide variation from pump to pump in the ease or difficulty with which argon instability occurs. In some cases it is necessary to feed the pumps argon in higher concentrations before they will go unstable.

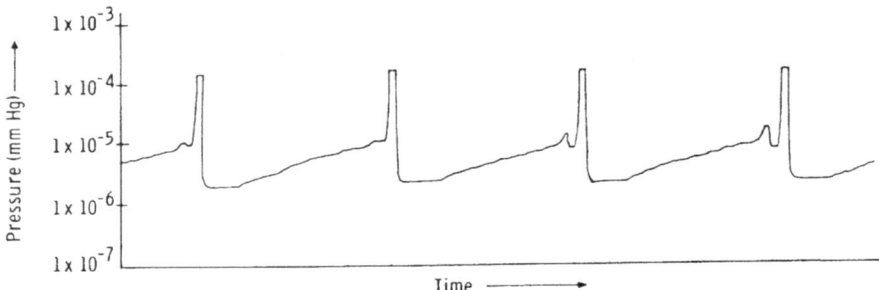

Fig. 5.4 Typical pattern of pressure vs. time for a getter-ion pump exhibiting argon instability (Jepsen et al., 1960) [4]. *Note*: 1 mmHg = 1 Torr = 133.3 Pa

- A representative pattern of argon instability is shown in Fig. 5.4. It should be remarked, however, that a considerable variety exists in the pressure vs. time curves that are observed.
- During an instability the pressure rises to a maximum value of about 2×10^{-4} Torr. For leak rates such that the pressure between instabilities is less than about 2 to 3×10^{-5} Torr, this maximum pressure is independent of leak rate.
- Time intervals between the pressure fluctuations are typically several minutes at a pressure of 1×10^{-5} Torr, and vary approximately inversely with leak rate for pressures in the range 5×10^{-7}–2×10^{-5} Torr.
- If the pressure exceeds about 3×10^{-5} Torr, instabilities no longer occur, even with pure argon
- The detailed nature of argon instability depends on such things as voltage–current characteristics of the pump power supply and size of system to which the pump is attached. On systems which are large relative to the size of the pump, the pressure fluctuations occur less readily because of "volume stabilization" of the pressure.
- Operation at sufficiently high pressures (as during the normal starting following rough pumping, for example) results in a substantial amount of pump cleaning and greatly reduces the tendency of a pump to exhibit argon instability. Thus pumps which are frequently cycled seldom go unstable.

Experimental Results [4]

From the numerous experiments thus far performed with VacIon pumps (Varian SIPs) employing slotted cathodes, useful cathode designs have evolved, and the following major results have been obtained:

- Air pumping: Slotted cathode pumps appear to be completely air stable. Speed is somewhat better than with smooth cathode pumps.
- Argon pumping: Pumping speed for argon is approximately 10 % of the speed for air. This is an improvement over smooth cathode pumps by a factor of about

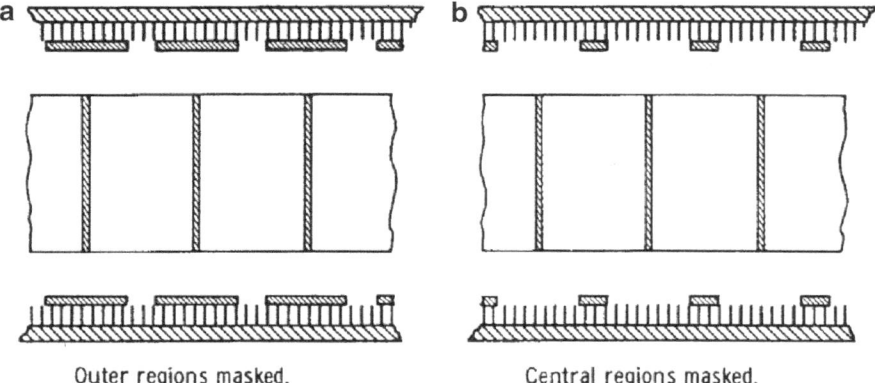

Outer regions masked.   Central regions masked.

Fig. 5.5 Crosssectional side of VacIon pumps employing slotted cathodes with (**a**) outer regions of the cathodes masked and (**b**) central regions masked [4]

five. Slotted cathode pumps appear to be stable on pure argon for pressures below 1×10^{-5} Torr. In most cases stable operation extends to even higher pressures.

- Cathode-limited life: A VacIon pump (Varian) with slotted cathode had operated stably on air for over 2,000 h at a pressure of 1×10^{-5} Torr. (This is equivalent, for example, to 20,000 h at 1×10^{-6} Torr.) Although noticeably worn, the slots were still effective in preventing argon instability. Other pumps operated at lower voltage and with greater conductance limitation have been run for the equivalent of 5,000 h at 1×10^{-5} Torr with comparable wear. These cathode-limited lives are appreciably less than those obtainable with smooth cathodes. They are, however, greater in general than the "flaking-limited" lives discussed below.

- Flaking: When deposits of material consisting of sputtered titanium and pumped gases build up to sufficient thickness, flaking may occur. When flakes fall into an area of intense discharge, some of the pumped gas may be released, causing pressure fluctuations. Flaking normally becomes significant after the equivalent of 1,000 to 5,000 h of operation at a pressure of 1×10^{-5} Torr, depending on the particular application and design of the pump. Subjecting the pump elements to a thorough cleaning restores the pump to "as-good-as-new" condition, except for the cathode material that has been consumed.

Upon examination of the slotted cathodes after extended operation, it was noted that in those portions of the cathodes located near the cell axes, where the most intense bombardment and most rapid erosion take place, little build-up of material occurs. It was therefore proposed that the faster pumping of argon takes place in areas outside the heavily sputtered central regions. To test this speculation, these two regions of the slotted cathodes were masked alternately, as shown schematically in Fig. 5.5.

Argon speed and stability behavior of the pump employing slotted cathodes with the *outer* portions masked were essentially the same as with completely smooth cathodes. With only the *central regions* masked, behavior appeared to be identical with that for the completely slotted cathodes.

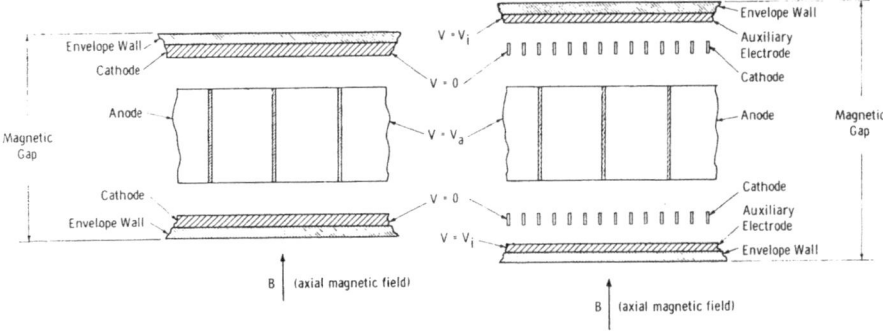

Fig. 5.6 Comparison of the magnetic gaps required for "equivalent" diode and triode pumps [4]

Since the slots are not necessary in the central regions, these areas can be left solid, thereby providing more material for sputtering and increasing life. In this way it may be possible to make slotted cathodes whose lives are comparable with those of smooth cathodes.

Jepsen et al. [4] compared the magnetic field efficiency in diode pump and triode pump [4].

Magnetic Field Efficiency [4]

One of the major cost factors in getter-ion pumps of the cold-cathode gas discharge variety is the magnet. It is therefore important to utilize the magnetic field efficiently. In a magnetic field of given intensity and volume, it is normally important to obtain maximum pumping speed for such active gases as H_2, N_2, O_2, and air.

Suppose that the magnetic field intensity and operating voltage (voltage between the anode and cathode) have been chosen. Corresponding to these parameters will be an optimum diameter for the individual anode cells. For a given set of magnetic gap dimensions there exists an optimum combination of anode length and cathode–anode spacing, as determined by considerations of "intrinsic" speed and conductance of gas into the anode.

An example of a diode pump of optimum design is shown schematically in Fig. 5.6. Shown also in Fig. 5.6 is an "equivalent" triode pump in which anode length, cathode–anode spacing, magnetic field strength, and voltage between anode and cathode are the same. In such a triode pump the magnetic gap must be lengthened because of the thickness of the collector electrodes, the spaces required between collector electrodes and cathodes, and the increased thickness of the cathodes themselves.

It was found experimentally that at intermediate pressures the pumping speeds for air of the two pumps are essentially identical. Since the volume of magnetic field occupied by the diode pump is less than that occupied by the "equivalent" triode pump, the pumping speed per unit volume of magnetic field is correspondingly greater for the diode pump.

5.1.5 *"Pumping Mechanisms for the Inert Gases in Diode Penning Pumps" by Baker and Laurenson [5]*

Baker and Laurenson [5] presented the important relationship as that the pump speed S measured in a dynamic system is dependent on the ratio R of the mean of the atomic masses of the cathode pair to the atomic mass of the impinging ion, in fact, $S \propto \log R$.

Pump Stability [5]

A subjective record of the behavior of each cathode pair was made whilst pumping a particular gas. For the purpose of the experiment pressures were kept constant by varying the leak rate to prevent as far as possible the instabilities which are common when pumping the inert gases. For instance during this work the Ti-Ti pump was reasonably stable whilst pumping argon, whereas given more time or by deliberately varying the leak rate, it would undoubtedly have exhibited cyclic pressure instability. Vaumoron and De Biasio [13] have noted that cyclic instability is observed when the ratio of the atomic mass of the "heavy" cathode in the cathode pair to the gas atomic mass is less than between 2.4 and 2.7. Results from this work put the ratio at between 2.2 and 2.4 if only the Ti-Ta and Ti-Ti pumps are considered. Intermittent pressure pulses rather than cyclic fluctuations were observed for the Al-Ta pump whilst pumping neon, argon, and krypton. During these pulses the pressure rose to about a hundred times its base value and fell again within the space of 1 min.

Pump Speeds [5]

The speeds of the different cathode pairs after each inert gas had been pumped for 4 h at 2×10^{-6} Torr are shown in Table 5.2. For argon the pumping speed was steady or increased slightly during the 4-h period but with the other gases the speeds decreased by an amount which was typically 20 % of the initial speed. There was also a general decrease of speed from the first to the second, to the third set of experiments. Figure 5.7 shows a log linear plot of the speeds versus the ratio R which is given by the average cathode atomic mass of the cathode pair divided by the gas atomic mass. The graph indicates that:

$$\text{Speed} \propto \log R,$$

so that, for example, the highest speed is observed with the "lightest" gas pumped by the "heaviest" cathode.

Our consideration:

The results suggest that for pumping a high-mass inert gas such as Kr, a high-mass cathode pair like Ti-Ta pair is essential to stable pumping.

Table 5.2 The ratio R, the stability and the dynamic speed during the 4-h pumping period are shown for the three cathode pairs and each gas [5]

Gas	Cathode pair	R	Stability*	Dynamic speed		
				First set	Second set	Third set
He	Ti-Ta	28.7	S	31	29	24
He	Al-Ta	26.0	S	26	23	...
He	Ti-Ti	12.0	S	...	...	22
Ne	Ti-Ta	5.73	S	16	15	13
Ne	Al-Ta	5.20	P	12	12	...
A	Ti-Ta	2.87	S	19	18	18
A	Al-Ta	2.60	P	13	16	...
Ne	Ti-Ti	2.40	S	...	...	9
Kr	Ti-Ta	1.36	F	...	10	11
Kr	Al-Ta	1.24	P	...	10	...
A	Ti-Ti	1.20	F	...	...	9
Kr	Ti-Ti	0.57	U	...	...	...

S stable, pressure almost constant, P intermittent pressure pulse, F small pressure fluctuations, U unstable, regular pressure fluctuations
*Stability during 4-h pimping period

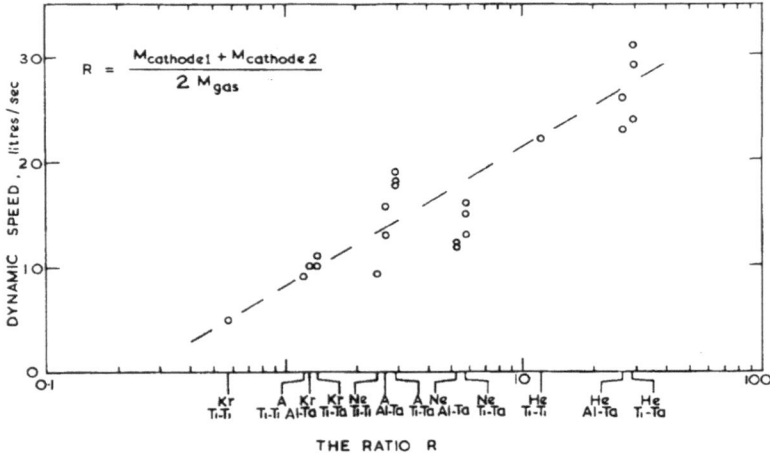

Fig. 5.7 The dynamic speeds of the three cathode pairs for each inert gas as a function of the ratio R [5]

5.2 Developing JEOL SIPs

We have developed two types of SIPs, one having the high pumping speed in the UHV range for UHV-EMs, and the other showing stable pumping performance for Ar gas, as well as high pumping speed in UHV range for AESs (Auger electron spectroscopes). We got the knowledge of SIP much by reading the instructive papers [1–5] introduced in Sect. 5.1. As a result, we could design and manufacture the desired two types of JEOL SIPs through experiments with reasonable cost.

We presented the experimental results of JEOL SIPs for extremely low pressure operation [14] and the pump with "slotted Ta on flat Ti/flat Ti" cathode pair dedicated for Ar gas pumping 1992 [15]. Here, designing and pumping characteristics of two types of JEOL SIPs, one having sufficient pumping speeds in UHV range and the other having stably pumping performance for Ar gas, as well as having sufficient pumping speeds in UHV range, are presented in the following Sections.

5.3 Designing and Pumping Characteristics of JEOL SIPs with High-Magnetic-Flux Densities (UHV-Type SIPs)

We have learned much with many papers, as mentioned in Sects. 5.1.1, 5.1.2, 5.1.3, 5.1.4, 5.1.5.

We discussed as follows.

- A magnet assembly of high-magnetic-flux densities must be necessary for extremely low pressure operation. Magnet assemblies with about 1.0 T were usually for many types of commercial SIPs. Magnet-flux density of 0.2 T might be upper limit for ferrite-made magnets. Rare-earth magnets are extremely high in commercial cost.
- Anode cells with relatively large diameters might be desirable for extremely low pressure operation. Anode cells with about 20 mm diameter were used for many types of commercial SIPs. We discussed that 1-in. (24-mm) SS304 pipe with thin wall are easily available.
- We must manufacture the JEOL-SIPs with reasonable cost. Make an experiment for searching best combination of magnetic-flux density and anode-cell diameter, considering reasonable manufacturing cost.

5.3.1 Pump Design Parameters

A common pump vessel of type-304 stainless steel (SS304) with two discharge chambers and with two high-voltage feedthroughs was made by vacuum brazing and Ar-arc welding. Dimensions of both discharge chambers were 58 mm × 250 mm × 100 mm. Three pump-element assemblies composed of two Ti plates (cathodes) and cylindrical anode cells (SS304) of different diameters, respectively, were designed and made, each of which was alternately assembled into the common pump vessel.

Magnet assemblies were provided:

#1; 0.15 T (ferrite)
#2; 0.2 T (ferrite)
#3; 0.3 T (rare-earth magnet)

Fig. 5.8 Distribution of
magnetic-flux densities in the
pole gaps of the magnet
assemblies: *Solid line*, y=0
mm; *dashed line*, y=30 mm;
open circle, 0.15 T at the
center of the pole gap; *cross*,
0.2 T; *filled circle*, 0.3 T [14]

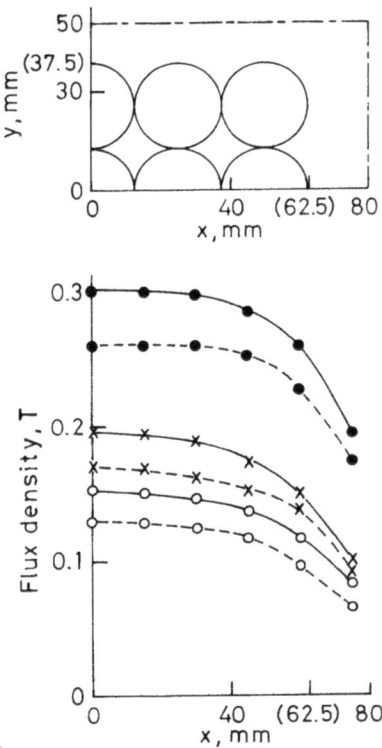

Anode cell assemblies were provided:

#1; 28 pieces with 17-mm-diam cells (4 columns × 7 rows)
#2; 15 pieces with 24-mm-diam cells (3 columns × 5 rows)
#3; 8 pieces with 29-mm-diam cells (2 columns × 4 rows)

The lengths of the anode cells of all the pump elements were 28 mm. Cathode
pairs of all the pump elements were composed of flat Ti plates of 2-mm thickness.
The gap space between the anode cells and the cathode plates was 10 mm. The
applied voltage between the anode and the cathode was 6.5 kV.

The magnetic-field profiles in the pole gaps of the respective magnet assemblies
were measured, and are presented in Fig. 5.8. The maximum flux densities at the
centers of their fields are just 0.15 T, slightly weaker than 0.2 T, and just 0.3 T,
respectively. The broken curves represent the flux densities at places 30 mm from
x-axis. For all magnet assemblies, the reduction of the flux densities at places
30 mm from the x-axis is about 13 %. The reduction of flux densities at places
60 mm from the y-axis is about 20 %.

Fig. 5.9 Experimental setup
for measuring pumping
speeds [14]

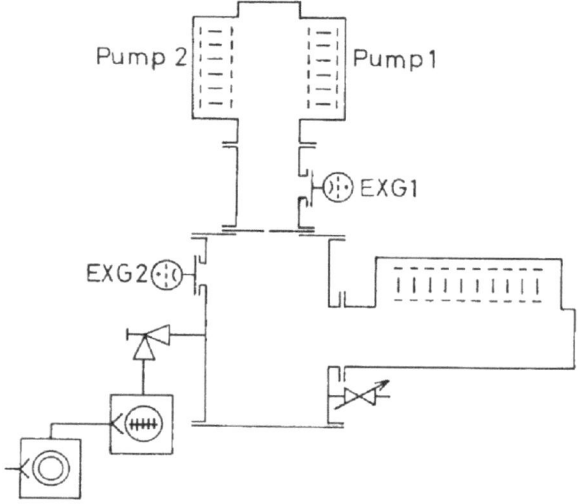

5.3.2   Pumping Speeds and Discharge Intensities

On measuring the pumping speeds the orifice (3.3 mm diameter) and the supple-
mentary SIP (75 L/s of nominal pumping speed) were installed as shown in Fig. 5.9.
The degassing treatment (250 °C for 2 days) was conducted for every pump, followed by the measurement of pumping speeds.

In measuring the pumping speed the pressure was first increased step by step
from a very low pressure of nearly 1×10^{-8} Pa to about 1×10^{-6} Pa by increasing the
introduced N_2 gas load through the needle valve, and the pressure was then reduced
step by step in the return course. After controlling the introduced gas load to set a
one-point pressure, the needle valve was not handled for about 10 min. The pumping speed measured in the pressure-fall course was about 10 % lower than the corresponding pumping speed measured in the pressure-rise course. The pressures in
the figures were thus measured at the end of these 10-min periods in the pressure-
rise course. Ion current characteristics were also measured at the same time in the
pressure-rise course.

The pumping speed characteristics and the discharge intensity I/P characteristics
of the pump with 17-mm diameter cells for N_2 are presented in Fig. 5.10, and those
of the pumps with 24- and 29-mm diameter cells are presented in Figs. 5.11 and
5.12, respectively.

The characteristics of the pump with 17-mm diam. cells in Fig. 5.10 show the following features. The discharge intensity characteristics indeed resemble the pumping
speed characteristics of the respective pumps. 1.0 A/Pa of discharge intensity corresponds to 10 L/s for the pumps with 17-mm diam. cells; however, there are some

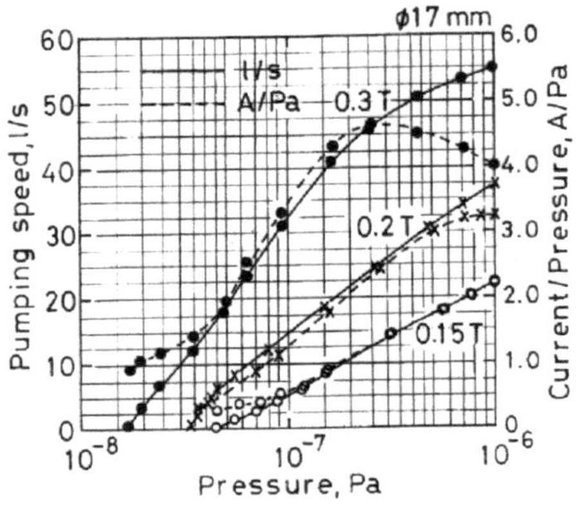

Fig. 5.10 Pumping speed
characteristics and discharge
intensity characteristics of the
pumps with 17-mm diam.
cells for N_2. *Solid line,*
pumping speed; *dashed line,*
discharge intensity; *open
circle*, 0.15 T; *cross*, 0.2 T;
filled circle, 0.3 T [14]

Fig. 5.11 Pumping speed
characteristics and discharge
intensity characteristics of the
pumps with 24-mm diam.
cells for N_2. *Solid line,*
pumping speed; *dashed line,*
discharge intensity; *open
circle*, 0.15 T; *cross*, 0.2 T;
filled circle, 0.3 T [14]

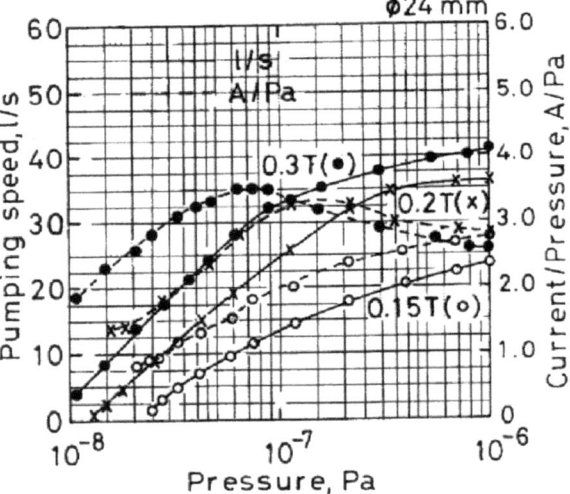

differences between both characteristics. That is, the discharge intensity of the pump
of 0.3 T again reduces with pressure rise when the pressure increases above
2.5×10^{-7} Pa, while the actual pumping speed increases with pressure rise up to
1×10^{-6} Pa.
The discharge intensity of the pumps with 17-mm diam. cells does not go down to
zero with pressure decrease, while the measured pumping speed goes to zero with
pressure decrease. This might be caused by outgassing from the pump elements.

The characteristics of the pump with 24-mm diam. cells in Fig. 5.11 show the
following features. Both the pumping speed and the discharge intensity of the pump
of 0.15 T increase smoothly with pressure rise up to 1×10^{-6} Pa. 1.0 A/Pa of discharge intensity corresponds to pumping speeds smaller than 10 L/s for the pump of

Fig. 5.12 Pumping speed
characteristics and discharge
intensity characteristics of the
pumps with 24-mm diam.
cells for N_2. *Solid line,*
pumping speed; *dashed line,*
discharge intensity; *open
circle*, 0.15 T; *cross*, 0.2 T;
filled circle, 0.3 T [14]

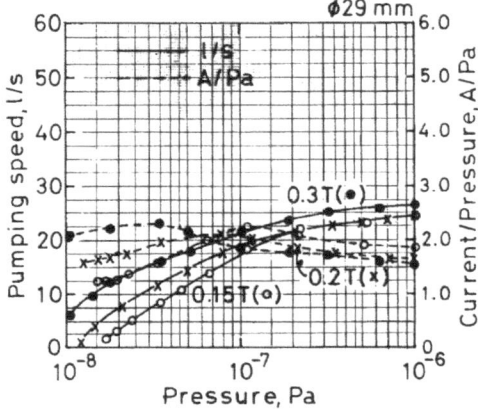

0.15 T. The pumping speed curves of the pump of 0.2 T crosses over the discharge
intensity curve at 2.4×10^{-7} Pa. In the pressure range below 4×10^{-7} Pa the pumping
speed of the pump of 0.2 T goes down with pressure decrease, and finally it becomes
almost zero at 1.3×10^{-8} Pa. The pumping-speed curve and the discharge-intensity
curve of the pump of 0.3 T cross each other at 1.2×10^{-7} Pa. The pump still shows a
considerable pumping speed in the 10^{-8} range, though the pumping speed reduces
with pressure decrease.

The characteristic curves of the pump with 29-mm diam. cells in Fig. 5.12 are
fairly different from those of the pumps of anode cells of smaller diameters. The
discharge intensity varies little with pressure in the 10^{-8} and 10^{-7} Pa ranges. On the
other hand, the pumping speed slightly reduces with pressure decrease.

There are the pressures at which the respective pumps show the maximum dis-
charge intensity, which are about 1×10^{-7} Pa for the pump of 0.15 T, 7×10^{-8} Pa for
0.2 T, and 3×10^{-8} Pa for 0.3 T, respectively.

Details of JEOL SIP with high-magnetic-flux densities were presented, which
is titled "Pumping characteristics of sputter ion pumps with high-magnetic-flux
densities in an ultrahigh vacuum range" by us 1992 [14].

5.3.3 How to Measure Very Small Ion Currents (Smaller than 0.1 μA) Precisely

Very small ion-currents of SIP must be measured for measuring its discharge intensities (I/P) in the UHV region. We measured such small currents by an analog-type
micro-ammeter directly by inserting the meter between the earth potential and the
controller-earth terminal by making the SIP controller isolated from the earth
potential.

Fig. 5.13 Distribution of
magnetic flux density in the
pole gap of the magnet
assembly [15]

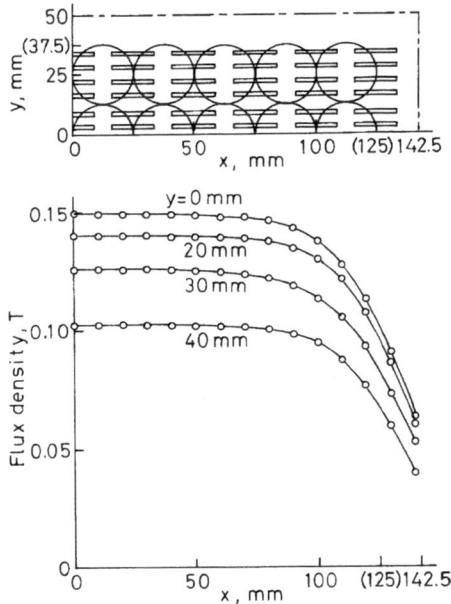

5.3.4   Designing and Ar-Pumping Characteristics of JEOL Noble SIPs with Various Shapes of 'Ta/Ti' Cathode Pairs

Various Shapes of Ti/Ta Cathode Pair

Pumping characteristics of SIPs depend on magnetic flux density, anode voltage, and dimensions of anode cells, especially the cell diameter. Further, Ar-pumping characteristics of diode-type pumps would be strongly related to the shape and material of their cathode pairs.

Four pump-element assemblies of different cathode pairs were replaceable with one another for the common pump vessel. The shape and dimensions of the anode cells were the same for all the pump-element assemblies. The anode-cell assembly was composed of 30 cells (24 mm diam., 28 mm length, 3 columns × 10 rows), as shown in the upper part of Fig. 5.13. The magnet assembly of 0.15 T at the center of the discharge area was made, whose dimensions were: gap space, 62 mm; magnet thickness, 30 mm; yoke thickness, 8 mm; magnet-pole area, 100 mm × 285 mm.

The distribution of magnet flux density in the pole gap was measured, as presented in Fig. 5.13. The maximum flux density at the center of the field is just 0.15 T. The reduction of the flux densities at the places 100 mm apart along the x axis from the center is about 7 %. The flux density higher than 0.1 T is measured at the places 40 mm apart along the y axis from the center.

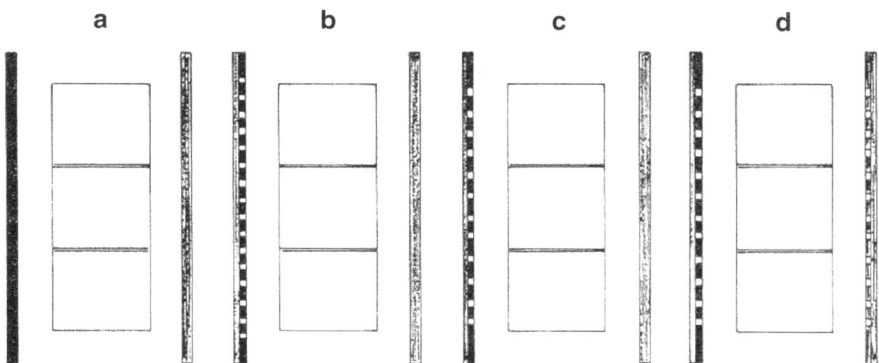

Fig. 5.14 Various shapes of Ta/Ti cathode pairs. (**a**) Flat Ta/flat Ti cathode pair, (**b**) holed Ta on flat Ti/ flat Ti pair, (**c**) slotted Ta on flat Ti/ flat Ti pair, and (**d**) slotted Ta on flat Ti/slotted Ti on flat Ti pair [15]

The four kinds of pump-element assemblies are presented in Fig. 5.14a–d. (a) "flat Ta/flat Ti" cathode pair, (b) "holed Ta on flat Ti/flat Ti" pair, and (c) and (d) "slotted Ta on flat Ti/flat Ti" pair and "slotted Ta on flat Ti/slotted Ti on flat Ti" pair, respectively. The arrangement of slots of slotted cathodes is seen in the upper part of Fig. 5.13. For slotted cathodes and holed ones, the center axis of all the anode cells face the flat surfaces, not slotted or holed area.

The average flux density over the discharge area was calculated as 0.13 T. The anode voltage was commonly set at 6.5 kV.

We measured the pumping speeds of SIPs with various shapes of "Ta/Ti" cathode pairs by an orifice method.

Ar-Pumping Speed Characteristics and N_2/Air-Pumping Speed Characteristics

In measuring Ar-gas pressures the relative sensitivity correction between two Bayard-Alpert gauges (BAGs) (one was a nude type and the other glass-tube type) was made. Ar pressures were obtained by multiplying a correction factor 0.71 (for Ar to N_2) to pressure indications. In measuring Ar-pumping speeds the pressure was increased from about 2×10^{-5} Pa up to about 4×10^{-4} Pa, step by step, by increasing the introduced gas load through a needle valve. After controlling the introduced gas load to set a one point pressure, the needle valve was not handled for 1–3 h. The pressures in the figures (Fig. 5.15) were thus obtained at the end of pressure-keeping periods, in the chamber to which the SIP was connected. Ion currents of the test pump were also measured at the same time in the pumping speed measurement.

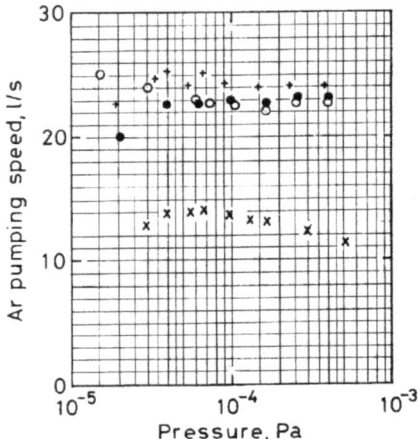

Fig. 5.15 Ar-pumping speeds of the pumps with various cathode pairs as a function of pressure. *Cross*, "flat Ta/flat Ti cathode" pair; *filled circle*, "holed Ta on flat Ti/flat Ti" pair; *open circle*, "slotted Ta on flat Ti/flat Ti" pair; *plus*, "slotted Ta on flat Ti/slotted Ti on flat Ti" pair [15]

Ar-pumping-speed characteristics of the pumps of the respective elements are presented in Fig. 5.15. The "saturated" Ar pumping speed of the pump of "flat Ta/flat Ti" cathode pair was 12–14 L/s in the 10^{-5} and 10^{-4} Pa ranges. On the other hand, the pumps of "holed Ta on flat Ti/flat Ti" pair and of "slotted Ta on flat Ti/flat Ti" pair have almost the same pumping speed as 22–23 L/s. The pumping speed of the pump of "slotted Ta on flat Ti/slotted Ti on flat Ti" pair was about 25 L/s, which is slightly larger than those of the pump of "slotted Ta on flat Ti/flat Ti" pair.

Ion current characteristics of all the test pumps were almost the same. Their curves in log-log plots in the 10^{-5} and 10^{-4} Pa ranges showed nearly straight lines inclining by 45°. Typical ion currents were, for instance, 0.26 mA at 2×10^{-4} Pa.

The pump of the "slotted Ta on flat Ti/ flat Ti" cathode pair was considered to be reasonable in manufacturing cost among three pumps with high Ar-pumping speed. The N_2-pumping speed and the air-pumping speed characteristics of this pump were measured as follows. In measuring the N_2-pumping speed the pressure was first increased from nearly 1×10^{-7} Pa to about 4×10^{-4} Pa, step by step, by increasing the introduced N_2 gas load through a needle valve. After controlling the introduced gas load to set a one-point pressure, the needle valve was not handle for about 20 min. The pressures in Fig. 5.16 were thus measured at the end of such 20-min periods. The air-pumping speed was also measured in the same way. The N_2-pumping speed and the air-pumping speed characteristics are presented in Fig. 5.16.

As seen in Fig. 5.16, the pumping speeds measured in the pressure-rise course are higher than the corresponding speeds measured in the returned course. Hysteresis for N_2 is stronger than that for air. The N_2-pumping speed in the pressure-rise course

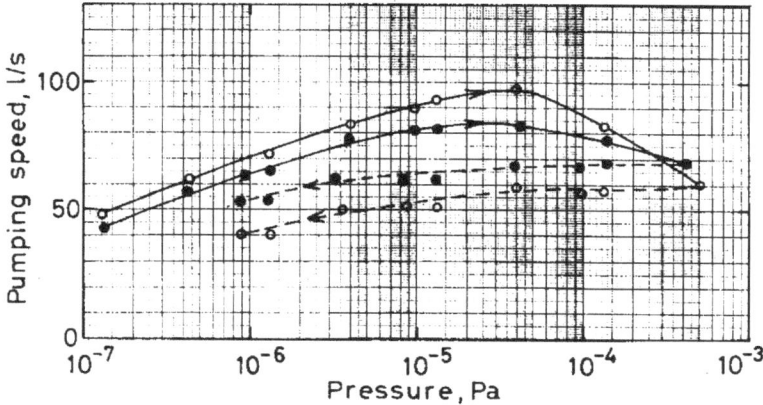

Fig. 5.16 N_2-pumping speeds (*open circle*) and air-pumping speeds (*filled circle*) of the pump of the "slotted Ta on flat Ti/flat Ti" cathode pair [15]

increased from 50 L/s at 1.3×10^{-7} Pa to 95 L/s in the 10^{-5} Pa range. The pumping speed in the return course gradually decreased with pressure decrease, for example 60 L/s at 5×10^{-4} Pa, 55 L/s at 1×10^{-5} Pa, and 40 L/s at 1×10^{-6} Pa. The air-pumping speed in the pressure-rise course increased gradually with pressure, from 43 L/s at 1.3×10^{-7} Pa to 85 L/s in the 10^{-5} Pa range. The air-pumping speed slightly decreased with pressure decrease in the returned course, for example 70 L/s at 4×10^{-4} Pa, 65 L/s at 1×10^{-5} Pa, and 55 L/s at 1×10^{-6} Pa.

The Ar-pumping speed of the pump of the "flat Ta/flat Ti" pair was 13 L/s in the 10^{-5} Pa range, and that of the pump of the "slotted Ta on flat Ti/flat Ti" pair was about 23 L/s in the 10^{-5} Pa range. The N_2-pumping speed of the pump of the "slotted Ta on flat Ti/flat Ti" pair was measured as about 50 L/s in the 10^{-6} Pa range and about 55 L/s in the 10^{-5} Pa range. Therefore, the Ar-pumping speed was as large as 42 % of the corresponding N_2-pumping speed of the pump of the "slotted Ta on flat Ti/flat Ti" pair. The pump of the "slotted Ta on flat Ti/slotted Ti on flat Ti" pair showed an Ar-pumping speed of 24–25 L/s, which were slightly larger than the speed of the pump of the "slotted Ta on flat Ti/flat Ti" pair in the same pressure range.

Details of JEOL noble SIP with high-magnetic-flux densities were presented by us 1992 [15], which is titled "Ar-pumping characteristics of diode-type sputter ion pumps with various shapes of 'Ta/Ti' cathode pairs"

Comments

1. The JEOL noble pump having "slotted Ta on flat Ti/flat Ti" cathode pair is used for JEOL JAMP-series Auger electron spectrometers (AESs). This JEOL noble pump has one 1 mm-thickness slotted Ta plate and one 1 mm-thickness flat Ti plate combined, and one 2 mm-thickness flat Ti plate for one pumping unit-housing,

resulting in cheep material cost. Besides, the slotted 1 mm-thickness Ta plate can be manufactured from the 1 mm-thickness Ta plate by the punching machine, resulting in cheep in manufacturing cost.

2. In order to pump Xe gas stably, the pump having "slotted Ta on flat Ti" cathodes for both sides is available, if desired. At the same time, it should be noted Ta plate is much expensive, compared with Ti plate.

3. 0.15 T of magnetic flux density may be the maximum limit for ferrite magnet in practical sense. When using the rare-earth magnet, 0.2 T or 0.3 T can be easily obtained. However, the cost of rare-earth magnet is much expensive. Additionally speaking, the ferrite magnet can be baked up to about 100 °C, on the other hand, the rare-earth magnet cannot withstand even mild-baking temperatures.

4. Comment: JEOL SIPs exclusively have 24 mm-diam. cells and 0.15 T magnet assembly.

5.3.5 Measuring the Pumping-Speeds of SIP Accurately by "Three-Point Pressure Method"

In the pipe method for measuring the pumping speed of the ultrahigh-vacuum pump such as SIP, the two-point pressure method for measuring the gas-flow rate through a conducting pipe is conventionally adopted, as presented by Munro and Tom [16]. In such two-point pressure method, the flow rate Q through the pipe is calculated as

$$Q = C_{12}(P_1 - P_2)$$

That is, the outgassing rate Q_W of the conducting-pipe wall is neglected.

However, in the actual system the introduced gas molecules are adsorbed and absorbed onto the conducting-pipe wall, and such absorbed gas molecules will desorbed during the period of the measurement of the pump speed, showing the distributed outgassing rates. Such distributed outgassing rates cannot be negligible for measuring the pumping speed in ultrahigh-vacuum region.

We reported "the three-point-pressure method (3PP method) for measuring the gas-flow rate through a conducting pipe" (1986) [17], and applied the 3PP method to measuring the pumping speed of an SIP. I introduce this new method by quoting some parts of the paper.

Principle [17]

Let us examine the pressure in a portion of a long outgassing pipe where the net gas flow including the leak rate Q_L directs to the left, as shown in Fig. 5.17. The conductance C of tube length L is calculated using the long tube formula assuming a linear conductance for a pipe portion.

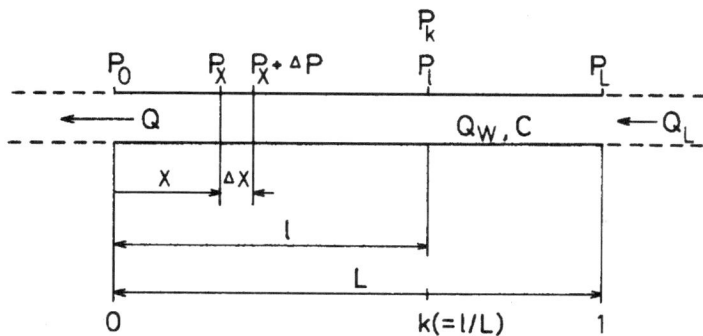

Fig. 5.17 Pipe portion in a long outgassing tube, where gas effectively flows from the *right* to the *left*. Gas-flow rate Q is composed of the leak rate Q_L and the outgassing rate Q_W [17]

Consider a very short element with a length Δx and a conductance $CL/\Delta x$. Since the total gas-flow rate at a position x is calculated as $Q_L + (1 - x/L)Q_W$, the pressure drop ΔP across the element is

$$\Delta P = \frac{\left[Q_L + (1 - x/L)Q_W\right]\Delta x}{CL}.$$

That is,

$$dP/dx = \frac{Q_L + (1 - x/L)Q_W}{CL}. \tag{5.3}$$

Integrating Eq. (5.3) from 0 to l with respect to x:

$$\int_{P_0}^{P_l} dP = \int_0^l \frac{Q_L + (1 - x/L)Q_W}{CL} dx$$

and so

$$P_l = P_0 + \left[Q_L + \{1 - l/(2L)\}Q_W\right]l/(CL),$$

where P_0 and P_l denote the pressures at the positions $x=0$ and l, respectively. Substituting a fraction k for l/L and rewriting P_l to P_k, the pressure P_k at a position indicated by the fraction k is given by

$$P_k = P_0 + \frac{k\{Q_L + (1 - k/2)Q_W\}}{C}. \tag{5.4}$$

Equation (5.4) shows that P_k varies quadratically as a function of k. Equation (5.4) contains three unknown factors Q_L, Q_W, and P_0. These three factors can be calculated from Eq. (5.4) using three pressures P_{k1}, P_{k2}, and P_{k3} measured at three different points in the pipe identified by their fractional positions k_i.

Gas-flow rates Q_L and Q_W are then expressed as

$$Q_L = C \frac{(2-k_2-k_3)(k_2-k_3)P_{k1}+(2-k_3-k_1)(k_3-k_1)P_{k2}+(2-k_1-k_2)(k_1-k_2)P_{k3}}{(k_2-k_3)k_2k_3+(k_3-k_1)k_3k_1+(k_1-k_2)k_1k_2}, \quad (5.5)$$

$$Q_W = -2C \frac{(k_2-k_3)P_{k1}+(k_3-k_1)P_{k2}+(k_1-k_2)P_{k3}}{(k_2-k_3)k_2k_3+(k_3-k_1)k_3k_1+(k_1-k_2)k_1k_2}, \quad (5.6)$$

where $0 \le k_i \le 1$. As a result, the total gas-flow rate $Q(=Q_L+Q_W)$ is expressed as

$$Q = -C \frac{(k_2^2-k_3^2)P_{k1}+(k_3^2-k_1^2)P_{k2}+(k_1^2-k_2^2)P_{k3}}{(k_2-k_3)k_2k_3+(k_3-k_1)k_3k_1+(k_1-k_2)k_1k_2}. \quad (5.7)$$

This is a good solution as long as the gauges are kept sufficiently far from the tube ends in the linear section. We call this gas-flow rate measuring method based on Eq. (5.7), the "three-point pressure method" (or simply 3PP method).

Process for Deriving Eqs. (5.5) and (5.6) [17]

The following equation in a matrix form is obtained for three pressures P_{k1}, P_{k2}, and P_{k3} in the pipe from Eq. (5.4):

$$\begin{pmatrix} P_{k1} \\ P_{k2} \\ P_{k3} \end{pmatrix} = \begin{pmatrix} 1 & k_1 & k_1(1-k_1/2) \\ 1 & k_2 & k_2(1-k_2/2) \\ 1 & k_3 & k_3(1-k_3/2) \end{pmatrix} \begin{pmatrix} P_0 \\ Q_L/C \\ Q_W/C \end{pmatrix}$$

Let $|\alpha|$, $|\beta|$, and $|\gamma|$ represent the following determinates:

$$|\alpha| = \begin{vmatrix} 1 & P_{k1} & k_1(1-k_1/2) \\ 1 & P_{k2} & k_2(1-k_2/2) \\ 1 & P_{k3} & k_3(1-k_3/2) \end{vmatrix},$$

$$|\beta| = \begin{vmatrix} 1 & k_1 & P_{k1} \\ 1 & k_2 & P_{k2} \\ 1 & k_3 & P_{k3} \end{vmatrix},$$

$$|\gamma| = \begin{vmatrix} 1 & k_1 & k_1(1-k_1/2) \\ 1 & k_2 & k_2(1-k_2/2) \\ 1 & k_3 & k_3(1-k_3/2) \end{vmatrix}.$$

Then, the values Q_L and Q_W are derived, using Cramer's formula, as $Q_L = C|\alpha|/|\gamma|$ and $Q_W = C|\beta|/|\gamma|$, which are the same as Eqs. (5.5) and (5.6), respectively.

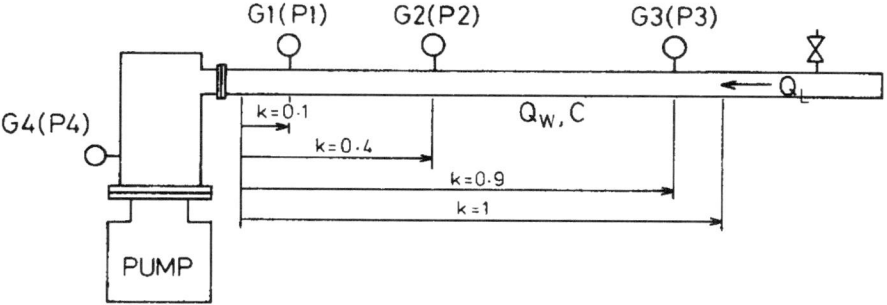

Fig. 5.18 Gas-flow rate measuring system with three gauges installed on a conducting pipe. G1, G2, G3, and G4 are B-A gauges; Q_L is the leak rate of the introduced gas; Q_W is the outgassing rate of the tube portion from $k = 0$ to 1 of the pipe; and C is the conductance of the same portion [17]

Optimization of the Measuring System [17]

For accurate calculations of Q_L and Q_W, the function k_1, k_2, and k_3 should be selected so that the differences between the pressures at the successive two points become nearly equal, i.e., $P_{k1} - p_{k2} \cong P_{k2} - p_{k3}$. As an example, when $Q_W<<Q_L$, the desirable value of k_2 is easily found to be 0.5 for $k_1=0.1$ and $k_3=0.9$. On the other hand, when $Q_L<<Q_W$, the desirable value of k_2 is calculated to be 0.36 using the following equation, which is reduced from Eq. (5.2) by neglecting the value Q_L:

$$P_k = P_0 + (1 - k/2)kQ_W / C.$$

In some cases, Q_W and Q_L would be comparable, so k_2 was actually selected as 0.4 for $k_2=0.1$ and $k_3=0.9$ in our measuring system.

Substituting 0.1, 0.4, and 0.9 for k_1, k_2, and k_3 of Eqs. (5.5)–(5.7), respectively,

$$Q_L = (5C / 12) \times (7P_{0.1} - 16P_{0.4} + 9P_{0.9}), \tag{5.8}$$

$$Q_W = (-5C / 3) \times (5P_{0.1} - 8P_{0.4} + 3P_{0.9}), \tag{5.9}$$

$$Q = (-5C / 12) \times (13P_{0.1} - 16P_{0.4} + 3P_{0.9}). \tag{5.10}$$

Measurement of Gas-Flow Rates [17]

A conducting pipe with three Bayard-Alpert gauges G1, G2, and G3 was connected to a typical test dome with an ion pump (60 L/s, triode type, Anelva). Three gauges were located at the positions indicated by the fractions k of 0.1, 0.4, and 0.9 as shown in Fig. 5.18.

A pipe of 43-mm diam. and 800-mm length (corresponding to the length L) with a conductance of 12 L/s was actually selected to make the pressures P_{01}, P_{04}, and P_{09} be in the same pressure range. The gauge G3 is located at a distance of 240 mm from the variable leak valve, so the ratio of the distance 240 mm to the diameter of the pipe 43 mm is as high as ~6. The ratio is considered to be enough for the gauge G3 being in the linear portion of the arrival rate distribution.

Substituting 12 for C and rewriting $P_{0.1}$, $P_{0.4}$, and $P_{0.9}$ to $P1$, $P2$, and $P3$, respectively, Eqs. (5–8), (5–9), and (5–10) become

$$Q_L = 5(7P1 - 16P2 + 9P3), \tag{5.11}$$

$$Q_W = -20(5P1 - 8P2 + 3P3), \tag{5.12}$$

$$Q = -5(13P1 - 16P2 + 3P3). \tag{5.13}$$

On the other hand, gas-flow rates are conventionally calculated using two pressures along the pipe. The gas-flow rate Q'_{1-3} calculated using the pressures $P1$ and $P3$ is given by

$$Q'_{1-3} = 15(P3 - P1) \tag{5.14}$$

where 15 is the conductance (L/s) between the gauges G1 and G3. We call this method the two-point-pressure method (or simply the 2PP method).

The following pretreatment was conducted:

- The system was baked under a high vacuum created by a turbo-molecular pump. The conducting pipe was "uniformly" baked at 200 °C for one day with the temperatures at three points of the pipe being controlled, during which the pump and the dome were baked at 300 °C.
- The electrodes of the ion pump were cleaned by an argon glow discharge with a current of 1 A at about 300 Vac for 20 min under a pressure of 40 Pa, during which the temperature of the pipe and the temperature s of the pump and the dome were kept at 150 and 200 °C, respectively.

Relatively accurate pressure measurements are very important to calculate the gas-flow rates by the 3PP method. Relative calibrations among the gauges were carried out for nitrogen at several pressure levels. Nitrogen was intermittently introduced into the closed system of Fig. 5.18 before the ion pump was switched on. The pressure readings of G1, G2, and G3 were recorded to correct their relative sensitivities at along elapsed time after introducing nitrogen. The sensitivity of the gauge G2 was 12 % higher than those of the gauges G1 and G3. On the other hand, the dependence of relative sensitivities of the gauges upon pressure levels was negligibly small. So the pressure readings of G2 were divided by 1.12 regardless of pressure levels.

After the system was evacuated to a pressure less than 1×10^{-6} Pa, nitrogen gas was leaked into the pipe through a variable leak valve to make the pressure $P1$ ~1×10^{-5} Pa. In the course of reaching equilibrium pressures, $P1$, $P2$, and $P3$ were read at elapsed times of 5, 15, 25 min. The leak rate was then increased to obtain higher pressures. The pressures $P1$, $P2$, and $P3$ were read at elapsed times of 5, 15, 25 min. The process was repeated at various pressures up to ~3×10^{-3} Pa.

The pressure distribution in the pipe at 25 min at two different pressure levels are shown in Fig. 5.19, where the solid curve shows the pressure distribution in the 10^{-5} Pa range and the broken curve in the 10^{-3} Pa range.

Fig. 5.19 Pressure
distributions in the pipe at an
elapsed time of 25 min for
nitrogen. *Filled circle* in the
10⁻⁵ Pa range, *open circle-* in
the 10⁻³ Pa range [17]

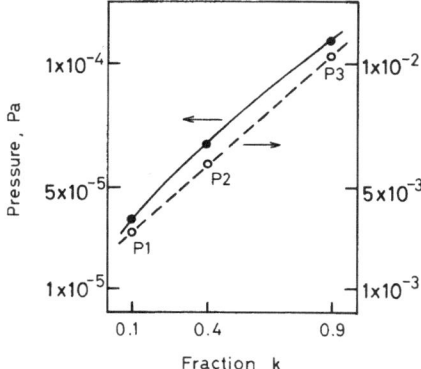

The convex curve in the 10^{-5} Pa range shows that the outgassing rate of the pipe wall is appreciable compared with the leak rate, while the straight line in the 10^{-3} Pa range shows that the effect of the outgassing is negligibly small.

Calculation of the Pumping Speed [17]

The pumping speed of the ion pump used can be calculated using the flow rate $Q(= Q_L + Q_W)$ and the dome pressure $P4$ measured by the gauge G4 (shown in Fig. 5.19) under the following assumptions:

1. The outgassing rate of the test dome is negligibly small compared with the rate of gases flowing into the test dome.
2. The ultimate pressure of the pump is negligibly low compared with the test pressure.

The pressure $P0$ and the pumping speed $S0$ at the pump mouth for nitrogen were calculated as

$$P0 = P4 - Q/1200,$$

$$S0 = Q/P0,$$

where 1200 is the conductance (L/s) between the pump mouth and the gauge G4, and Q is the gas-flow rate given by Eq. (5.13).

The pumping speed $S0'$ is also calculated conventionally using Q'_{1-3} of Eq. (5.14) (the conventional pipe method).

Pumping speeds for nitrogen as a function of pressure at an elapsed time of 25 min are presented in Fig. 5.20, where the solid line curve shows the speed $S0$ derived using the value Q calculated by 3PP method and the broken line curve $S0'$ derived using the value Q'_{1-3} by the conventional pipe method [17].

The 3PP method can give real gas-flow rate when three pressures are measured accurately.

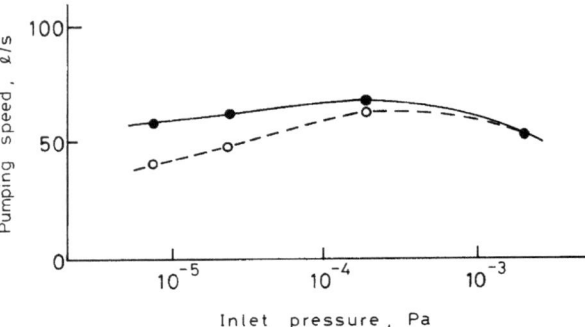

Fig. 5.20 Pumping speeds for nitrogen as a function of pressure at an elapsed time of 25 min. *Filled circle* $S0$ by the 3PP method, *open circle* $S0'$ by the conventional pipe method [17]

5.4   Know-How Technology Found in Degassing SIPs

Manufacturing process for SIPs is as follows.

Machining to form metal parts
Cleaning the machined metal parts
Welding to form the pump vessel and others
Degassing the pump itself

5.4.1   Degassing Process for SIPs

Step-1: Several pump vessels are connected to an ultrahigh-vacuum chamber using copper pipes to be evacuated by the turbo-molecular pump. And ascertain there is no air-leakage in the degassing vacuum system using a quadrupole mass-filter.

Step-2: The pump vessels are subjected to degassing treatment (around 200 °C) for several days. A furnace, or tape heaters are used for baking. Then pump vessels are gradually cooled to around 50 °C.

Step-3: The magnet assemblies are attached to the pump vessels and the SIPs are switched "ON" to be activated. And the working SIPs are subjected to mild baking (60–80 °C) for several days.

Step-4: The power for baking was switched "OFF" and the SIPs are gradually cooled to room temperature, keeping SIPs being "ON". The SIPs are to be operating for several days, and ascertain the extremely low pressure by the pump current and the extractor gauge.

5.4.2   Know-How Technology

* Among the processes listed above, Step-3 is indeed effective for degassing the interior of the SIP, which is due to electron/ion stimulated desorption (ESD/ISD). The effect of ESD and ISD is described in details in the book written by Yoshimura [18].

- In Step-3 listed above, introduction of a very small amount of pure Ar gas, to increase the saturated pressure to 10^{-5}–10^{-6} Pa, is very effective to clean the interior of SIPs due to being covered with sputtered Ti molecules. However, the process of Ar gas leaking could be omitted in actual degassing process to make the process simple.Additionally speaking, I ascertained the effect of pumping a small amount of Ar gas by SIP as follows: I vented the column of a JEM equipped with a large SIP in our laboratory with Ar gas for about one minute, and then the microscope column was routinely evacuated according to automatic sequence. I measured the pump-down characteristics. The pressure was surely pumped down faster, compared with the case of N_2 venting. And the saturated ultimate pressure was lower than the pressure just before this experiment, which was due to the activated SIP due to Ar-gas pumping.
- Regeneration of the SIP whose interior is oxidized:Parts such as the anode-cell assembly and the cathode plates of Ti or Ta can be regenerated by removing the oxide surfaces with the glass-shot/steel shot blasting. However, Pump vessels cannot be thoroughly cleaned by the glass-shot/steel shot blasting or acid cleaning. Pump vessels and ceramic parts such as the electric-current feed-through and insulators must be exchanged.

References

1. Jepsen RL (1968) The physics of sputter-ion pumps. In: Procedings of the 4th international vacuum congress, 1968, Manchester, pp 317–324, April 1968
2. Rutherford SL (1964) Sputter-ion pumps for low pressure operation. In: Transactions of the 10th national vacuum symposium, 1963, Macmillan, New York, pp 185–190
3. Andrew D (1968) The development of sputter-ion pumps. In: Proceedings of the 4th international vacuum congress, 1968, Manchester, pp 325–331, April 1968
4. Jepsen RL, Francis AB, Rutherford SL, Kietzmann BE (1961) Stabilized air pumping with diode type getter-ion pumps. In: Transactions of the 7th national vacuum symposium, 1960. Pergamon, New York, pp 45–50
5. Baker PN, Laurenson L (1972) Pumping mechanisms for the inert gases in diode Penning pumps. J Vac Sci Technol 9(1):375–379
6. Jepsen RL (1961) J Appl Phys 32:2619–2626
7. Kornelsen EV (1964) Can J Phys 42:364–381
8. Winters HF, Kay E (1967) J Appl Phys 38:3928–3934
9. Tom T, James BD (1966) Abs. 13th AVS vacuum symposium, pp 21–22
10. Brubaker WM (1960) Transactions of the 6th AVS vacuum symposium, 1959. Pergamon, New York, pp 302–306
11. Bance UR, Craig RD (1966) Vacuum 16:647–652
12. Carter G (1963) Transactions of the 9th AVS vacuum symposium, 1959. Pergamon, New York, pp 302–306
13. Vaumoron JA, De Biasio MP (1970) Vacuum 20:109
14. Ohara K, Ando I, Yoshimura N (1992) Pumping characteristics of sputter ion pumps with high-magnetic-flux densities in an ultrahigh vacuum range. J Vac Sci Technol A 10(5): 3340–3343
15. Yoshimura N, Ohara K, Ando I, Hirano H (1992) Ar-pumping characteristics of diode-type sputter ion pumps with various shapes of 'Ta/Ti' cathode pairs. J Vac Sci Technol A 10(3):553–555

16. Munro DF, Tom T (1965) Speed measuring of ion getter pumps by the 'three-gauge' method. In: 1965 Transactions of the 3rd international vacuum congress, pp 377–380
17. Hirano H, Yoshimura N (1986) A three-point-pressure method for measuring the gas-flow rate through a conducting pipe. J Vac Sci Technol A 4(6):2526–2530
18. Yoshimura N (2008) Vacuum technology: practice for scientific instruments. Springer, Berlin
19. Hamilton AR (1962) Transactions. of the 8th AVS vacuum symposium, 1961. Pergamon, New York, pp 388–394
20. Andrew D, Sethna D, Weston G (1968) Proceedings of the 4th international vacuum congress, pp 337–340

Chapter 6
Ultrahigh Vacuum Electron Microscopes

Abstract Field-emission (FE) emitters are indispensable for atomic resolution EMs (ARMs) and analytical EMs (AEMs). Therefore, we had to do best to produce UHV in EMs on market.

6.1 Field-Emission Electron Sources

6.1.1 Electron Sources

Y. Harada and Y. Tomita made some comments on the course of history of electron emission sources 2011 [1].

Development of Electron Sources [1]

The following electron sources have been used for electron microscopes:

Thermal electron sources; W hairpin filament, W point filament, and LaB_6 emitter

FE sources; W<111> or W<100>T-FE (thermal FE) source, W<310>C-FE (cold FE) source, ZrO/W<100> (Schottky emission) source.

They are optionally used according to performance, life, vacuum requirement and cost. Their characteristics are given in Table 6.1.

Kim et al. [2] compared an FE cathode and Schottky emission (SE) cathode, as follows.

Table 6.1 Typical electron emission source [1]

	Thermal W-hairpin	Thermal LaB$_6$	T-FE ZrO/W < 100>	C-FE W < 100>
Brightness at 200 kV(A/cm²/sr)	~10^5	~10^6	~10^8	~10^9
Spot size	~20 µm	~10 µm	~20 nm	~10 nm
Energy width (eV)	~3	~2	~0.8	~0.3
Operation temperature (K)	~2,800	~1,800	~1,800	~300
Operation pressure (Pa)	~10^{-4}	~10^{-5}	~10^{-7}	~10^{-8}
Emission current (µA)	~100	~30	~100	~10

Table 6.2 Comparison of cold field emission tip and Schottky emission tip [2]

Property	Cold FE source	SE source
Cathode	W single crystal <111>, <310>	Zr/O/W
Radius	100 nm	About 1 µm
Temperature	300 K	1,800 K
Work function	4.4 eV	2.7 eV
Virtual source size	~5 nm	~15 nm
Emission current	~10 µA	~100 µA
Angular emission	<100 µA/sr	>100 µA/sr
Energy spread	0.3–0.8 eV	0.6–1.2 eV
Heating power	none	~2 W
Stability (noise)	A few %	<1 %
Vacuum pressure	10^{-10}–0^{-11} Torr	10^{-8}–10^{-9} Torr

Cold FE Source and SE Source [2]

Table 6.2 shows the comparison between cold FE and SE sources. The cold FE cathodes typically used are single crystal <111> and <310> tungsten (W). Most cold FE cathodes have a radius of about 100 nm after annealing. The SE cathode most commonly used is a ZrO coated <100> oriented tungsten emitter. The ZrO reduces the work function of W <100> from 4.5 to 2.7 eV. A disadvantage of a SE emitter is that the tip operates at 1,800 K and requires a few watts of heating power. It also has a larger energy spread compared to a C-FE emitter at a given angular emission-current density. However, SE emitters have a much more stable emission current than the C-FE source due to the lower electric fields and the larger and more stable emitting areas. Typically, it has less than 1 % of noise fluctuation over several hours. The SE emitter can be operated under poorer vacuum conditions than the cold FE source. The operating pressure range for the SE emitter is 10^{-8}–10^{-9} Torr [2].

6.1.2   How to Check the W<310> FE Emitter

The field emitter W<310> was applied to the SEM first for JEOL microscopes. We tried to measure the extremely low pressures using an extractor gauge in the gun chamber evacuated with a JEOL SIP.

Fig. 6.1 Typical dependence
of emission current with time.
As the tip becomes coated
with contaminants, the
emission first of all drops,
and then begins a steady rise
until the emission becomes
erratic and the tip will
eventually destroy itself with
a vacuum arc [3]

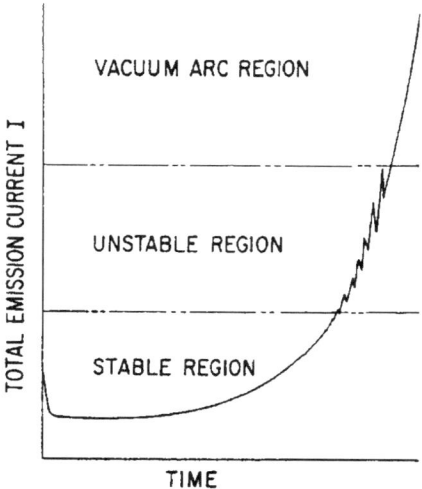

An etched W<310> tip was welded on the W-filament. We called it simply the W<310> tip here. The process for formation of FE emitter is almost same as that presented by Crewe [3], as follows.

Tip Operation [3]

The etched W<310> tip is mounted in an enclosure which is evacuated to about 10^{-9} Torr. The tip is "formed" (rounded off) by sending a brief pulse of current through the filament. As the magnitude of the current pulse is increased, the filament begins to glow red. By this time the heat has usually driven off the contaminants so that the pressure no longer rises during the flash. The tip is then tested to see whether cold field emission is occurring. After emission is detected, a check is made to determine if the tip is properly formed by comparing an experimental voltage–current curve with that predicted by the Fowler-Nordheim equation, or not.

The subsequent performance of the tip appears to be more dependent on the local gas pressure than on any other parameter. In general, the emission current at constant voltage appears to be a curve similar to that of Fig. 6.1. At first, there is a small decline in emission current as the surface of the tip becomes coated with contaminants which increase the work function. Thereafter the current rises until it becomes erratic, and the tip eventually destroys itself by a vacuum arc. The time scale for this process can vary from seconds to thousands of hours depending on the pressure.

Our process to form the tip was almost same as the process in "Tip operation" cited above [3]. The period of "stable region" in Fig. 6.1 is a measure of the vacuum degree of very low pressure. Considering these situation, we try to compare the indications of the extractor gauge (EG), the ion current of the JEOL SIP used, and the period of the stable region of the total emission current.

First, we degassed the EG by an electron bombardment. And, following the long evacuation by the SIP (1.5 kG, 1-in. anode diameter), the characteristics of total emission current, like as shown in Fig. 6.1, were measured. Indication of the ion current of the SIP was almost zero for the very low pressure. Indication of the EG was ~10^{-8} Pa.

Next, we switched "OFF" the EG, and several 10 min later the characteristics of total emission current were measured. Then, the period of the stable region of the total emission current became longer comparing to the period during the EG being "ON." This means that the EG functions as an outgassing source.

Next, we switched "OFF" the SIP with keeping the EG "ON." Then, the ion current of EG rose gradually. When the ion current of the EG rose one order, we switched "ON" the SIP again. Then, the ion current of EG reduced rapidly, indicating an extreme low pressure of 10^{-9} Pa range. This checking showed that the JEOL SIP works well in the extremely low pressure range.

We discussed and reached the conclusion that the EG cannot be used for checking the pressure in the gun chamber. We concluded that the pressure around the FE emitter should be judged from the period of the stable region given in Fig. 6.1.

6.2   Progress Toward UHV Evacuation Systems with Sputter Ion Pumps

Philips (Holland) introduced the EM with a large sputter ion pump (SIP) on the market early 1980s. Recently, TEMs with several sputter ion pumps (SIPs) have become widely used.

SIPs are suitable for UHV scientific instruments because they can be baked up to about 150 °C for degassing. SIPs do not have any moving mechanism, so they are most suitable to electron microscopes (TEM and SEM) and Auger electron spectrometer (AES). SIP with a sufficient pumping speed in extremely low pressure region is required for the field emission (FE) gun chamber.

We presented a commentary article [4] on the development of JEOL UHV EMs 1987. Three types of the evacuation systems of standard JEOL TEMs are schematically shown in Fig. 6.2.

"DP cascade system" in (a) has been described in details in Sect. 4.1.6. A bypass valve of small conductance is omitted to draw. In the "Dry pumping system 1" in (b), the viewing chamber can be evacuated by DP in series system, as same as the DP in-series system of the SEM, presented in Sect. 4.1.7. In the "Dry pumping system 2" in (c), the viewing chamber is evacuated by a turbo-molecular pump (TMP). When an anti-vibration damper is used for a low-vibration type TMP, the vibration due to TMP can be isolated from the viewing chamber.

6.2.1   Examples of UHV Evacuation Systems for JEMs

The evacuation system of a typical UHV TEM, JEM-2000FXV, was presented in Fig. 6.3.

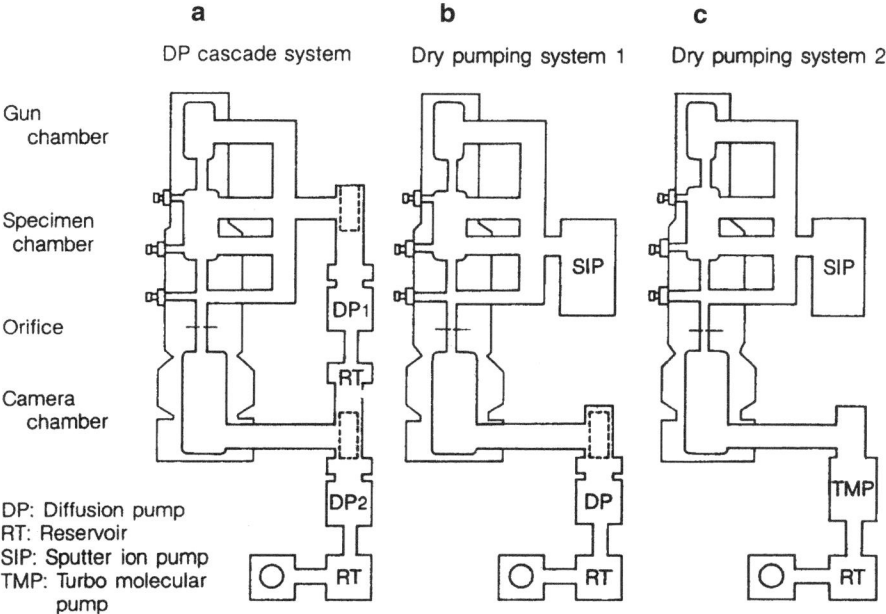

Fig. 6.2 Three types of evacuation systems of standard TEMs (JEOL TEM) [4]

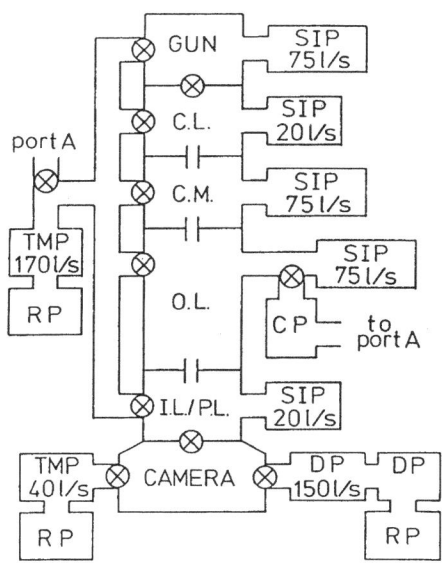

Fig. 6.3 Vacuum system of UHV-TEM, JEM-2000FXV [4]

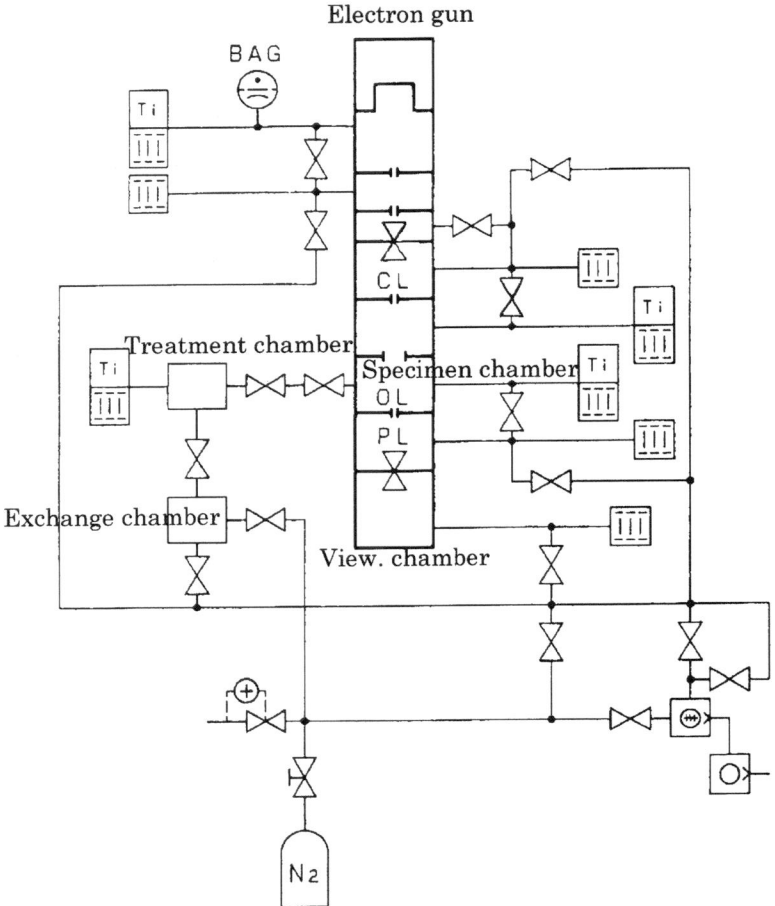

Fig. 6.4 Vacuum system diagram of a JEOL UHV STEM [5]

Two TMPs (170 and 40 L/s) is used for roughing the microscope column and the viewing chamber, respectively. After roughing, the inside of the TMPs and RPs are vented after switching OFF the pumps in order to stop the vibration of TMPs. The cryo-pump (CP) for the objective lens space is switched OFF to stop the vibration of CP during the period of observing and taking photographs the EM images.

Tomita [5] reported on the evacuation system of an UHV scanning TEM (UHV STEM) and its vacuum characteristics. A W<100> thermal FE (TFE) emitter is used in the UHV STEM, whose vacuum system diagram is presented in Fig. 6.4.

Key technologies for the UHV EM-column are metal-gasket sealing for the entire column and baking the column at up to 150 °C under high vacuum. Metal O-rings are used for sealing the electron-optics (EO) lens column and chambers in

Fig. 6.5 Block diagram of vacuum system. Microscope column is separated into six chambers, electron gun (GUN), condenser lens (CL), condenser mini-lens (CM), objective lens (OL), intermediate and projector lens (IL/PL) and camera (CAMERA) chambers. Individual vacuum pumps are attached to each chamber. *PEC* specimen pre-evacuation chamber, *PEG* Penning gauge, *PIG* Pirani gauge, *BAG* B-A gauge, *CP* Cryogenic pump [6]

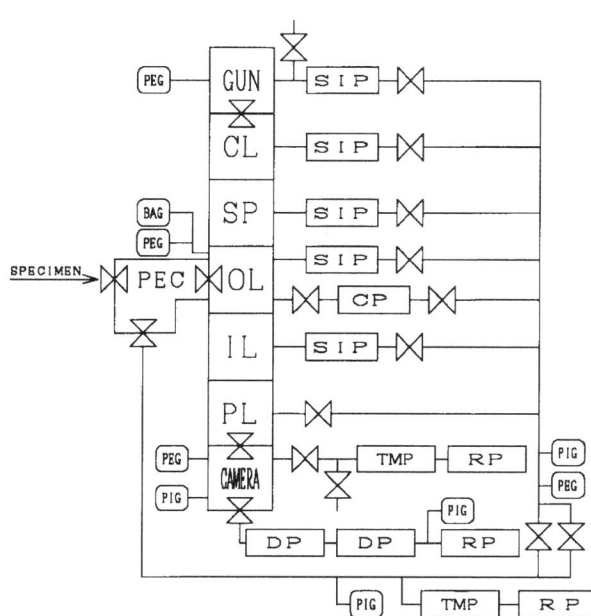

order to achieve assembling accuracy. On the other hand, vacuum parts such as valves and gauges are sealed using ICF (ConFlat®) flanges. The EO lens column, the specimen chamber and the gun chamber of the ultrahigh-vacuum STEM can be baked up to 160 °C [5].

Kondo et al. [6] described the 200 kV UHV high-resolution TEM (UHV HRTEM) for in situ surface observation.

The block diagram of the vacuum system of the UHV TEM/REM is presented in Fig. 6.5.

As seen in the Figs. 6.3, 6.4, 6.5, TMPs, backed by RPs, are used as the roughing pumps for respective parts of the microscope column. Such TMP-RP roughing systems are usually switched OFF and vented with dry nitrogen gas except roughing period.

In the UHV evacuation system of Fig. 6.4, it is noted that the viewing chamber is evacuated by a SIP in fine pumping. In the UHV evacuation systems of Figs. 6.3 and 6.5, it is noted that the viewing chamber (same as the camera chamber) is evacuated by a DP–DP in-series system in fine pumping.

References

1. Harada Y, Tomita M (2011) Development of technology of electron microscope in Japan. Microscopy 46(Suppl 3):2–47 (in Japanese)
2. Kim HS, Kratschmer E, Yu ML, Thomson MGR, Chang THP (1994) Evaluation of Zr/O/W Schottky emitters for microcolumn applications. J Vac Sci Technol B 12(6):3413–3417

3. Crewe AV, Eggenberger DN, Wall J, Welter LM (1968) Electron gun using a field-emission source. Rev Sci Instrum 39(4):576–583
4. Harada Y, Yoshimura N (1987) Significance of vacuum technology in electron micro-scope. J Vac Soc Jpn 30(12):985–988 (in Japanese)
5. Tomita T (1990) Ultrahigh-vacuum scanning electron microscope. Electron Microsc 25(2): 114–117 (in Japanese)
6. Kondo Y, Ohi K, Ishibashi Y, Hirano H, Harada Y, Takayanagi K, Tanishiro Y, Kobayashi K, Yagi K (1991) Design and development of an ultrahigh vacuum high-resolution transmission electron microscope. Ultramicroscopy 35(2):111–118

Afterword

It is natural that the electron microscope (EM) is provided with a field emission gun (FEG) for atomic resolution. For the analytical EM, a thermal FEG with high current density is naturally required. Also, a UHV specimen chamber is required for contamination-free atomic resolution observation. Given these requirements, we need two types of sputter ion pump (SIP), one for extremely low-pressure operation and the other for pumping argon or xenon gas for specimen thinning. We successfully developed two types of SIP for JEOL EMs (JEMs). I believe that UHV technology supports the analytical electron microscope (AEM) with an extremely high resolution of 0.1 nm.

This book describes the JEOL UHV technology for JEMs. I am confident that the technology described in this book will serve well for many kinds of scientific instruments because many of them require silent UHV pumps at a reasonable cost and clean UHV evacuation systems with safety systems so that no errors will occur in controlling the vacuum system. There are many kinds of scientific instruments requiring a UHV environment.

Many readers are very much interested in electron microscope technology itself. The monograph "Progress of electron microscope technology in Japan" was issued as a supplement to the journal *Microscopy* [Ref. [1] of Chap. 2], and contains many epoch-making TEM/STEM image photographs. The monograph cites a surprising number of references—as many as 297 papers and monographs. The English version of the "Progress of electron microscope technology in Japan" would be very useful for engineers and researchers worldwide.

As the author of the present volume, I am very grateful to the Microscopy Society of Japan for making it possible to include two figures: Fig. 2.1 and Fig. 2.7 ([1] in Chap. 2).

About the Author

Nagamitsu Yoshimura

1965: Graduated from Osaka Prefecture University, Engineering Division

1965: Entered JEOL Ltd

1985: Received Doctor of Engineering degree from Osaka Prefecture University. Ph.D. Thesis: "Research and Development of the High-Vacuum System of Electron Microscopes" (in Japanese)

1995: Qualified as a consultant engineer in physical field by passing the qualifying examination

1965–2002: Engaged in research and development of vacuum-related technology in electron microscopes for more than 35 years at JEOL-group companies

2002: Retired from JEOL Ltd

N. Yoshimura, *Historical Evolution Toward Achieving Ultrahigh Vacuum in JEOL Electron Microscopes*, SpringerBriefs in Applied Sciences and Technology, DOI 10.1007/978-4-431-54448-7, © The Author(s) 2014